THE INTELLECTIVE SPACE

Cary Wolfe, Series Editor

(continued on page 173)

THE
INTELLECTIVE
SPACE

Thinking beyond Cognition

Laurent Dubreuil

posthumanities 32

University of Minnesota Press
Minneapolis • London

Published by the University of Minnesota Press
111 Third Avenue South, Suite 290
Minneapolis, MN 55401-2520
http://www.upress.umn.edu

Library of Congress Cataloging-in-Publication Data

Dubreuil, Laurent.
The intellective space : thinking beyond cognition / Laurent Dubreuil.
(Posthumanities ; 32)
    Includes bibliographical references and index.
    ISBN 978-0-8166-9480-8 (hc : alk. paper)
    ISBN 978-0-8166-9485-3 (pb : alk. paper)
1. Thought and thinking. 2. Cognitive science. 3. Cognition. I. Title.
    B105.T54D83   2015
    128'.3—dc23

                                                        2014019920

Printed in the United States of America on acid-free paper

The University of Minnesota is an equal-opportunity
educator and employer.

21 20 19 18 17 16 15   10 9 8 7 6 5 4 3 2 1

Contents

I

THE INTELLECTIVE HYPOTHESIS

COGITATION AND COGNITION

1.

We say more than we think; we think more than we say. This does not sum up all of our lives, but, at least, it describes where we are now, you and me, and where we stand each time we reflect on something or exchange ideas and signs. This strange place, I call it *the intellective space,* that is, a putative space where thought and knowledge are performed and shared, and not only computed according to universal laws that would "speak" to us directly and by themselves.

All the ideas that we have seem to be supported somewhere, by a screen, a page, a sound, a tablet, an electric flux, matter or energy. As far as we know, we need, at some point, something like a brain to take charge of such ideas. The contemporary word of *cognition* generally refers to the minimal level of these mental operations. "Our" ideas, then, are both the effect and the product of cognitive actions. I suggest there is an *excess* of (and to) cognition, because of the material channels it has to go through—human verbal language, neural synapses, social prescription . . . Although this excess is often neutralized, it never completely disappears, and it shapes our thoughts more than once. The intellective is a possible name for the productive undoing of cognition per se. It points toward the potential journey of ideas going *beyond* cognition, after and before computation—this latter operation paving the way for virtual intellection, again.

Cognitive sciences are apt to be established, though not "intellective" ones. Indeed, the *institution* of science relies on an assumption that is favorable to pure (or, more realistically, *purified*) cognition. Therefore, a cognitive study systematically integrates a phraseological dimension, and its success goes with self-justification for the administration of knowledge. But it would be hasty to dismiss the very existence of cognition (or the merit of its scientific examination) based on such an intricacy of goals, for this type of situation is far from unique in the realm of *scientia.* Traditionally, the "intellective disciplines" have been more discursive and more attuned to the "humanities." Paradoxically, most practitioners of those latter fields would be unable to acknowledge this very fact and unwilling to make the first efforts to grasp the cognitive. So it is quite patent that future intellective studies, as structurally nonscientific as they might appear, would need to be immersed in a scientific milieu, even if their regular locus is in the humanities. At any rate, the division between thinking and its institution is not easy to determine, which invites us to consider in detail the conditions of our enunciation.

The introduction and circulation of words into animal cognition consolidates the advent of the intellective. With words as semantically unstable agents of relative conceptual stability, we may get more than actualization, mistake, and recognition. We say less than we think, for human verbal language forces us to focus, to discriminate. We say more than we think, when we hear in our speech something added to the impetuous ideas we once thought we would say. We are no longer in the exclusive domain of equality; we enter a space governed by more *and* less.[1]

2.

As the computer was being gradually discarded as an insufficient *model* for the functioning of the human brain, the tremendous multiplication of personal PCs and other laptops reinforced the value of the *analogy,* among people who had little familiarity with neuroscience. More polemically, one could also observe that our minds are being forcefully "rewired" by the mechanics of cell phones, social networks, and omnipresent softwares and that, in a matter of years, our brains might be reconfigured by our own computers, magically restoring the validity of good old cognitivism.

Aside from this bitter remark, it is manifest that the downfall of the computer as a model, combined with the vivid afterlife of the conjecture under an analogical guise, did not ruin the prestige of the "Turing machine." Many prominent scholars still abide by the doctrine, while distancing themselves from the computer under its current form and keeping the essential Machine. To some extent, they may be well-advised to do so. Willy-nilly, the basic mental operations that Alan Turing outlines are presupposed within the scientific presentation of the cognitive. As soon as one ratifies this category, a *complete* rejection of the mechanical mind is mainly for show, and it would be made largely inaudible anyway by the computerization of contemporary society. Accordingly, I consider cognition to be the *automaticity of mental activities*. With an accurate comprehension of the substrate of thought, of its settings, and of its rules (these are aspects we are just *beginning* to grasp, and they differ considerably from our current computers), we are—or will be—able to yield some statistically valid predictions. We could announce the probability of a process "occurring by itself" (something *auto-matic*), comparable to the formation of a molecular body made of different atoms in a given context. But, again, we are far from this prophetic moment, which, ultimately, would not give us access to the whole of thinking. Furthermore, the computer is one very limited example of the extent of the automatic, which additionally implies a capacity toward self-organization.

Disconnected from the computer, computability keeps its "analogical" relevance, insofar as we move away from oversimplified computations with ones and zeros (the *basic* image of one isolated neuron being excited or not) and concede the possibility of extremely complex arrangements, whose equations are largely obscure to us so far. And here, I "add" that thought is also *performed,* which is much more difficult (impossible?) to calculate.

3.

We are constitutively different. Our particular genes condition our brains, and their expression also depends on random circumstances. Development made us. My age, my mood, my food, my drinks and medications are apt to modify my neural functions. The thousands of experiments in brain imagery show one thing: variability affects our thoughts. No two individuals have the exact same embodied brain, nor do they use it in the exact same manner. I, too, in my own thinking, will differ from myself.

Across comparable populations, patterns of similarity do appear, and of course, this is all the more distinct with the help of a mathematical projection of "the-human-brain" (a few different models are currently in use). Notwithstanding extreme variability, some functional universals exist. They are often overrated, both for epistemic reasons (the scientific emphasis on the general) and more contingent ones (owing to the kinds of algorithms that are employed, to the techniques of imagery, to a lack of statistical corrections, to the social uniformity of the subjects, etc.), but I have no intention to deny their reality. A logical temptation is simply to *discount* everything that is not common, and this is basically where mainstream cognitive neuroscience situates itself today. In the meantime, lots of humanists continue to celebrate a quite standardized concept of *diversity.*

Variability is not the full extent of what there is, and it is not a marginal error either. We do not think *in spite of* it, or *against* it, as some traditional approaches borrowed from psychology and philosophy would lead us to admit; we just think *with* it: every one of us, and, much more obviously, all of us together, when we try to coordinate ourselves through any intellectual exchange. Discarding the singular is obeying the technical approach to communication as it was developed after World War II: getting rid of *noise.*

Claude Shannon, when he introduced his diagram of communication (Figure 1),[2] wrote plainly that his hypothesis solved an "engineering problem" (in the telephone industry and elsewhere) and that the "meaning" of the signal was outside its scope (or "irrelevant"[3]). Shannon even complained, a decade later, about what he perceived to be undue extrapolations of his theory.[4] For instance, one could meditate on the strong will to adapt an engineering model to literary studies (with the mediation of Roman Jakobson and the development of semiotics in the 1950s and onward), despite the fact that, in literature, the issue of meaning is supposed to be supreme. And it is not difficult to recognize that, in the sciences, this defining moment was also trying to diminish the part of meaning (and especially verbal semantics) in the description of mental activities. Here and there, we can observe a recourse to the technical as the best theoretical solution, allied with both the wild proliferation of one hypothesis and a stereotyping of the object of inquiry. Such underlying assumptions are as disputable as they are historically dated.

Even if we suspend the reservations we just outlined, an immediate problem is tied to the identification of *noise* with the noncommon. Deciding in those terms derives from a statistical approach,

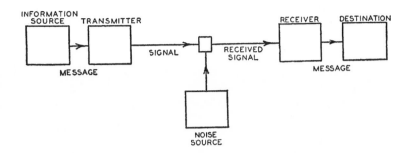

Figure 1. Schematic of a general communications system. From Claude E. Shannon and Warren Weaver, *The Mathematical Theory of Communication*, Figure 1. Copyright 1949, 1998 by the Board of Trustees of the University of Illinois. Reprinted with permission of the University of Illinois Press.

or a conceptual prescription consisting in *correcting* errors. Moreover, the noise of *noēsis* may make sense to us, that is, influence cognition. The erasure of variability could be required for the approximate depiction of basic cognitive function; thereafter, it cannot lead to any sound description of thinking.

4.

All that crosses our minds requires time to come and go. The event-related potentials, which were first recorded on electroencephalograms in the late 1920s and 1930s, are a common measure of responses and actions in the central nervous system. A more difficult task, a lack of attention, or some dysfunctions will create delays. Much of what is mentally processed happens in milliseconds, giving us the impression of the quasi-instantaneous. There always remains a physical timeline as well as some kind of lag that it would be quite relevant to call *différance*. The term of *cognition* itself, as its suffix indicates, alludes to a necessary unfolding.

We need to take the full measure of these remarks if we want to build up the notion of the intellective. The well-known paradox of the *now* (as soon as I point out the present moment, it is past) is inscribed in our nerves. The impossibility of the present is generally suspended in the self-constituting effectuation of thought, which conversely operates in the now: a recollection identifies a scene as belonging to yesterday, on the condition that it is being processed now in our memory. One could interpret this whole conundrum as the sign of a structural mental illusion, as an amusing but minor enigma, as a proof that we're living under the regime of *quasi*. I maintain that even the intimately perceived solidity of the *nunc cogitans* is irremediably fractured though practically real. A gradual succession of instants is no solution, because it just implies a relative refection of the discrete through punctual accumulation.

Hence, temporal cognition is a flux, *and* it is fundamentally broken. This is the first performance of its operation. Cognitive sciences tend to assume experimentally an addition of units, each one of them being complete, that is, attached to a measured time t_n. Concurrently, we are well aware of the almost relentless activity of the brain and of its plasticity. This discrepancy had been largely explored, decades before the emergence of the new sciences of the mind, by authors such as Henri Bergson. Bergson opted in favor of what he saw as the instinctual dynamic of life, whose verbal and theoretical expression had to remain deceptive (or mainly evocative) because of the analytic qualities of human intelligence and language. This is a brilliant proposition, though it is grounded in too many disputable suppositions (such as the élan vital or the large independence of thought vis-à-vis the brain). Moreover, we do not have to choose. If cognition constantly appears as two distinct ways (or more), then we should presume it is so. Cognition is developed (or *is*) through its interruption. Here differences between the living and the material, the instinctual and the intelligent, the animal and the human, are superfluous. Then, by isolating two different planes (instead of, say, dimensions), one loses the singularity of thought. Thinking is one and the other, *at the same time.* This *same time* (which cannot be the same, though it has to be so) is an opening toward the intellective.

5.

The modern description of processual thought has been changed by the introduction of multiple polarities. René Descartes, in his second *Meditation,* is exclusively concerned with the interaction of two poles, *cogito* and *cogitatio*. Descartes's insistence on the site of thought (*ego cogito,* or "I think") is of utmost importance. In many respects, the Cartesian doctrine is at odds with contemporary scientific materialism. Still, his "I" as *a point of articulation from which thought is being thought* remains a crucial requirement to cognitive studies—if, in accordance with the original text, we do not identify the Cartesian *I* with "the concept of the self" but with *something* involved in and defined by the activity of thinking (the *res cogitans,* or "thinking thing"), and if, in contradistinction with the philosopher (who defined the material body as *res extensa,* or "extended thing"), we tolerate an essential *extent* of thought.

So we have in Descartes a *cogito* producing (and produced by) *cogitations*. The *Meditations* also briefly refer to a "*vis cogitandum*"[5]—"a force" or "a will to think"—which is not directly defined but is a gift from God; and to "*modes* of thought" (*modi cogitandi),* including perception, a category that Spinoza will later amplify. All in all, the duality *cogito–cogitatio* will continue to command the dominant understanding of these issues. In the last century, Edmund Husserl, in his rereading of the *Meditations,* developed the *cogitatum,* or "what is thought,"[6] as a term relativizing Descartes's methodic solipsism. In each *cogito,* there is something that is turned into thought by what I *am,* as the insuppressible subjective sentiment of something external would illustrate. A further bifurcation is possible, between that which is thought as the result of cogitation and what is to be thought. In *Difference and Repetition,* Gilles Deleuze explored the interplay of *cogitatum* (what is thought) and *cogitandum* (what has to be thought),[7] aptly noticing the violent resistance of the unthought to its thinkability. Besides its etymological link to the vocabulary of the *cogito,* the *cognitio* (or "cognition") of cognitive sciences is first and foremost concerned with the study of this capacity of transformation.

In this basic scientific model, a *cogitation* has a *cogito* for locus (such as the embodied and synaptic brain), which is the receptacle for the content of thought. *Cogitandum* and *cogitatum* are, respectively, an *input* or an *output* (worst-case scenario), or, better, the index of something *external* to thinking and the transmissible *content* of thought. *Cognition* would refer to the principles of organization permitting the appearance and sustenance of *cogito* and/

or to the effective trajectory from *cogitandum* to *cogitatum*. That cognition could be all of this *at the same time* echoes ambiguities we already encountered. A temptation is to consider cognition as a coding machine, or as the code itself. In this case, we would be very close to uncover the nth repetition of the communication model. Besides, what is our exact interest in discovering *cognition* if we are able to emit *cogitata* from an *I* to another? The traditional answers are that cracking the code would help us grasp the reality of the real, understand each other straightforwardly through the reduction of cognitive deformations, create machines that would think as we do and become our interlocutors. What would we gain from direct access, as long as cognition has to take place? And why should artificial intelligence be *similar* to our own minds? I am not even addressing feasibility. Finally, we are left with unfinished business. The equivocation of *cognition* (as a set of rules, and/or a circumscribed transcription operation, and/or an episodic process of emergence) diffracts cogitation. Thinking is affixed to its locus but not contained within it. It is not a result *(cogitatum)* so much as a dynamic *(cognitio)*. As for the unity of its process, it seems to be at fault. In short, some reconfiguration is urgent.

I would rather maintain that cognition literally makes us think, and, in the particular enactment of common rules, it shapes what is identified as thought. My *cogito* (and all that pertains to the complexity of my *I*) influences my cognitive structures and, subsequently, its content. In turn, the *cogitatum,* as soon as it is formed, goes through the process of being thought again. Whereas *metacognition* is a reflexive system of control, thinking of one's thought *(meta-cogitation)* is somehow indirect. What is thought potentially modifies the *cogitandum.* Under this last word, I am doing my best to designate what is to be thought, which is the part of reality on the verge of becoming our object. The broken link between the *cogitatum* and the *cogitandum* does not imply that the unrelated *(le sans-lien)* is an effect of our thoughts. To a certain extent, I subscribe to a critique of correlationism. I rather aim to inscribe that what is to be thought by us is apt to be apprehended through the prism of what has already been thought. A rejuvenated Sapir–Whorf hypothesis and many suggestions of social biases informing our perceptions could be placed here. The *cogitandum* is deliberately a mixed notion, posited as a point of entry and a possible arrival, as an abrupt condition and a heteronomous entity. Whatever real reality could be, we are embedded within *other* cogitations as soon as we are born and, most definitely, before this. What is to be thought is, in this sense, both foreign to us and already *marked*. But a mark is not a sign. Within the *cogitandum* is to be found the index for both the unthinkable of thought (something independent from what I think is acknowledgeable, at the price of making its radical autonomy dependent of my own thought) and its post hoc (what is to be thought now also comes to me as already constituted). The whole circuit of cogitation is looped in a complicated way, with no systematically identical predetermined paths, which expands the role of neural reentry that has been rightly emphasized by Gerald Edelman. His model of "reentrant cortical integration"[8] in the brain finds an *analogon* in the process of thinking I just gave and is key for an anatomical understanding of the mental phenomena on which I will dwell.

6.

Throughout this book, affects are considered to be a part of thinking. In the same way I will refuse the dichotomy "thought *vs.* language," I do not recognize the traditional division between cognition and emotion. We may have to deal with distinct experiences and categories, but the usual separations are not acceptable. My speech belongs to what I think, and of course, my concepts are also affective. Not only "feelings" are produced by the mind, along with perceptions, but it is simplistic to situate them *exclusively* in the "lower" (subcortical) regions of the brain and/or the unconscious—whereas "authentic" cognitive operations would be the attribute of consciousness in prefrontal areas of the neocortex. I need affects to understand language or the face of the other. I also reflect on my own feelings and construct *sentiments* in my embodied thought. There are swift and rather automatic affects, inducing stereotypical responses, such as the basis for fear, anger, or love, and they are to be found in other animals than humans. The amygdala is clearly involved in the circuits of reward, sexual arousal, defensive behavior, or fear and anxiety in vertebrates, including mammals. Conditioning, as well as some associated behaviors (stress, aggression, submission), is real in all animals. But we luckily have surpassed the approach of Pavlov, Skinner, or Laborit: emotions cannot be reduced to or identified with the automatic responses of which behaviorism was so fond, or to their chemicophysical support in the central nervous system. Affects are also complex mental events, and they are much more than the passionate noise perturbing the glorious rule of pure reason.

The performance of cognition is irremediably affective. We might want to keep some distinctions, as long as they do not serve to exclude, out of principle, an intellectual mode from *noēsis,* and argue in particular that emotions are "subpar," or "negligible," or "corrigible" for actual thought. This is the old (and still vibrant) conception of impassible thinking, concealing both the irregularity of knowledge and its performance. Damasio was certainly well-advised to appeal to Spinoza against the dis-affected version of cognition. But, for Spinoza, the affect of the mind is still a "confused idea,"[9] and, I'm afraid, its anchorage in the particularities of the body is too restrictive for our purpose. We might better argue for a *differential integration* of affects (as feelings, emotions, or sentiments) and perceptions into the very process of thinking, a hypothesis apparently corresponding to the modes of cerebral projections and connections.

7.

I know that, by abandoning a more traditional apodictic form *(more geometrico),* I am making myself less "readable" by people I also would like to convince. With some work, everybody is able to understand the discursive sinuosity of thought, in the same way mathematical formalization could be acquired. Reaching a high level of proficiency, being able to create are, of course, much more difficult tasks, requiring time, focus, dedication. As for the almost spontaneous anxiety of the "professionalized philosopher" toward the density of an argumentative text written beyond the boundaries of the hypotheticodeductive, this is nothing more than the disciplinary effect of usage. If a scientist similarly were to say of the present text *I cannot make anything out of it,* this would be both expected and incorrect. Certainly this book is not *readily* applicable, not *readily* teachable. It is not *ready* to be delivered, for the present needs to be unpacked.

We are used to a disconnect between the experience of knowledge and its results. In writing as I do, I am simply putting the emphasis on the (personal and transindividual) ways through which we come to know what we seem to know. I am less driven by the "infinite possibilities of language" than by a desire to restrain the confidence in transparency. Moreover, there is a tangential confusion between the *incomprehensible* and the *"incompressible"* (to use a term derived from the algorithmic information theory Kolmogorov and others began constituting in the 1960s). The complexity of thinking lies in a resistance to the abbreviating procedures of summaries and descriptions. While an argument is certainly less "complex" in this respect than a perfectly random association of words (if it were), the repeated appeal to stylistic simplicity as a large compression of ideas is just a phraseological stance, provoked by social and political settings—and not by the "inner demands" of an intellectual task.

Standardized communication is readier for two main reasons: it is expressed within conventions that have been inculcated over years of domestication through schools and *habitus,* and it tries— modulo poetic neutralization—to limit the discursive volatility of thinking. The conduct of science undoubtedly benefits from it, in touching its *general* point of utterance, a point preliminary to the large reproducibility of both the formal and the experimental. But scientific *creation* is not there; it does not lie in the ideal of mere repetition. It needs unconventional, singular appropriations and distortions, which the experience of knowing encompasses. Refusing to be *ready* means refusing to be read like a disk or a tape.

We may *engineer* thought. And we may want to do more.

A HOLED FABRIC

8.

The *intellection* is the variable and processual performance of *cognition*. The *cognitive* is the retroactive enclosure of what I think on the operation of its cognition. It is usually presented as being non-affective and as the *terminus ad quem* of rationalism. The *intellective* stems from the extension of cognition beyond itself.

9.

We have two problems. One is scientific: how could we accommodate the role of variability and noise in cognition, enacted as a temporal process? or, Do we have an appropriate model for intellection? The other is about the limits of the cognitive and, as such, those of common rationality and scientific inquiry. Even a better biological and mathematical representation of performed cognition will leave us at the border of what lies beyond. Our course in the intellective space is impossible to calculate, but the gates we pass through are somehow determinable, and they influence our journey. One could hope that probabilities (especially if they are Bayesian) have some heuristic power. They might help our own *approximate* anticipation of a course, even though we should not forget that the probable is neither virtual nor actual. The accuracy of probability is strengthened by the fact that most of our mental experience is like a *cabotage,* where social norms, psychological habits, and epistemic neutralization tend to draw us back close to the shores. It ensues that, especially if they modify their fundamentals, cognitive sciences deliver a relatively correct model of our most crucial mental ability, especially when they are lower order. But they are inherently incomplete and would have no excuse to promote their lacunary description as political prescription.

Whereas the scientific problem is not for me to solve, it is at the core of our interrogation. The most current hypotheses consider thought as a complex dynamical system. The first equations on such systems date from the pioneering work of Henri Poincaré.

Later on, René Thom's "theory of catastrophes" was close to the point, on a qualitative (and philosophical) level. That said, the larger development occurred over the last three decades. The import of complexity and chaos into the study of cognition has been pretty much in progress since the 1980s. Thinking, envisaged either as a rapidly changing process involving an "astronomic" number of neurons (and neural connections) or as a tremendously flexible exchange of words, affects, and ideas between individuals, with all the hidden parameters implied therein, seems to be a perfect case for research into complex dynamical systems. The intersubjective aspect, the fragmentation of phenomena, and the multiplicity of factors are obvious to many scholars of philosophy and "theory," so the promises of a science of catastrophes, chaos, or complexity elicited many hopes among "humanists," from Michel Serres or Gilles Deleuze to Isabelle Stengers or N. Katherine Hayles. This is not the main path that has been taken by scientists, who rather focused on the dynamics of cortical networks or of ideation. Yet the two levels (micro and macro) are evidently interconnected, because trans-individual cognition has to be performed by brainlike structures. Furthermore, a distinctive feature of complexity theories is their scale-invariant applicability.

Complex dynamical systems have been introduced in physics in an effort to move away from the restrictive linearity famously advocated by Laplace. One tries to include the dense interaction of elements over time and, in particular, the deformation of the preceding value once it is reused through iterations (then, in a nonlinear fashion). In these cases, depending on the value of a variable c, the behavior of a graph is apt to be no longer regular but stochastic, with "movements" that are *not* predictable. It is an illustration of the role that "initial conditions" or "noise" are able to play, even in an ideal and controlled setting. Furthermore, in physical systems, noise is also introduced through measurement. The slightest approximation in determining x (it does not need to be a reading error) is extraordinarily amplified by successive iterations, which rapidly creates a dire divergence between the model and the system (the "butterfly effect" of Edward Lorenz).

When a large number of nonlinear elements interact with each other over time, under the right conditions, emergence could be exhibited at some point. The vast and interrelated association of separate entities suddenly creates a transient and unexpected pattern. *Emergence* threatens to become a magic word, as more and more cognitive functions are being qualified as such. It sometimes retains a kind of religious or mystical connotation. It remains one of the most plausible explanations. Authors such as J. A. Scott Kelso or Dante Chialvo advance the argument that cortical networks form a complex dynamical system, that noise and the stochastic activity of neurons do command the actual enactment of cognition, and that, consequently, thought is emergent. I do not intend to repeat their demonstrations. What I want to underline is this: in such conjectures, emergence indicates that cognitive acts are a combination of blocks, but also *more than that*. In my own vocabulary, this insight already relates to the *excessive* performance of cognition.

10.

Here is a fable. As in quadratic maps, we would have an alternation of chaotic and periodic behaviors in a dialogue: the ideas we are emitting, reusing, exchanging, and rearranging are parts of a complex dynamical system. In thinking together, we are making iterations, we are including the performance of our interlocutors within our own, and not just receiving or replicating it. Were this a cognitive model for our thoughts, determining rather constant and common mental operations across agents, sharing performance would still make the urge for retrieving the *original x* quite vain, and the scientific prediction rather undecided. But we might find some points of stabilization of the graph (or strange attractors around which the discussion would tend to oscillate), some possible zones of convergence in the *stream of consciousness,* the ones that are "public" or those standing *between* us.

Unexpectedly, we ex-change our thoughts among ourselves.

11.

The revised and provisional version of the scientific problem posed by intellection leaves us with a wide array of reflections. It first appears that the categories in use are somehow divided, perhaps split open. The formalization of *dynamics* is anchored in a kind of stasis. Nonlinearity is coming from discrete iterations, while the form of the equation itself is doomed to be progressive rather than processual and to focus on the output rather than on the course. In its turn, the definition of the *system* may be more rigid than it should: in particular, are its boundaries fixed, or are they dynamical as such? The *noise*, in becoming absolutely crucial in the development of these complex arrangements, is also converted into something other than itself: "noise, the only possible source of new patterns,"[10] once wrote Gregory Bateson. If it is no longer marginal or corrigible, but something without which there would be no emergence, is it still noisy? Materialism is ingrained in cognitive science, so it justifies the recourse to hypotheses borrowed from physics. But thought, once seen as an emergent property of the brain, appears to be less "material" than expected.

We are also touching the impossibilities of the scientific project through the means it granted to itself. The unpredictability of chaotic behaviors is not an outcome of an insufficient theory; it is the very consequence of the model, whose validity is experienced through its inability. As for measurement, its accuracy is not at fault when its inevitable approximation engenders the incommensurable of the real with respect to its mathematical encryption. On both counts, the usual claim of science is at odds with its own trajectory. This could be a perhaps regrettable though inexorable difficulty if we were in quantum physics, but such an observation carries even more weight for the physics of the mind. What is experienced here is simply the intellective, and it goes over the limits of the cognitive configuring the institution of science.

Dynamics, noise, and matter could be what they are, and differing from this *at the same time*. The meaninglessness of the negligible could be meaningful. The unpredictable, the incommensurable could be the negative resistances of the predictive model or measurement itself. And they are concurrently the affirmation of a surplus, or what "emerges" from emergence. The words I wrote might be unacceptable in the pure space of rationality. It just happened that we are somewhere else now, in the journey of the intellective.

12.

Kurt Gödel's "incompleteness theorems" triggered countless comments, appropriations, and interpretations since their first publication in the early 1930s. I am not presenting a new reading, but I cannot omit the connections between the epistemic situation we face and the proofs Gödel obtained against the possibility of complete formalism. In a very simplified way, the first theorem establishes the existence of a statement within Peano arithmetic (a consistent formal theory) that is neither provable nor refutable, that is, undecidable, in this theory. The second theorem states that number theory has no internal mean to prove its own consistency, unless it contradicts itself (and is inconsistent). It is usually admitted that Gödel's demonstration applies to mathematics in general and shows the formal impossibility for formalism to found itself. A more general (and quite accessible) formulation of the two theorems is, in following mathematician Jean-Yves Girard, (1) if a theory $\mathscr{T}$ is "sufficiently expressive" and consistent, then there is a formula G that is true but not provable in $\mathscr{T}$, and (2) if $\mathscr{T}$ is "sufficiently expressive" and consistent, then it does not prove its own consistency.[11]

This fundamental incompleteness, I believe, does not belong to mathematical formalism exclusively. By analogy, it is possible to *extend* the range of the second theorem beyond its scientific expression. The specificity of mathematics lies in its ability to code and prove its own incapacity. But the defect Gödel designated in a special case is sensible throughout thinking. Each system of knowledge is, within the limits of its own theory and rules, "incomplete." It cannot decide of all of its own statements. And it is led to contradiction every time it attempts to *overrule* its incompleteness. Knowledge is inconsistent in virtue of its configuration and logic; its fabric is holed. Formalization is a powerful way of keeping the sys-tem of the cognitive together, though it fails more absolutely (according to its own standards). Even the formal has to be performed and is susceptible to the intellective extension of cognition.

13.

Knowledge comes to us through discipline(s). At an existential level, *discipline* implies training, control, and, yes, some pain. "The disciplines" are organized sets of mental habits, methodic beliefs, expert practices, collective rites, modes of research, and referential traditions. A discipline, both a social body and an individually acquired *manner,* channels knowledge. What a discipline entails in terms of separation, relative stability, and collective sanction fosters the apparition, transmission, and accumulation of the known through the act of knowing. Our cognitive system needs focus, and it also benefits from the prostheses provided by textual canons, mathematical operations, or common assumptions. Disciplinary boundaries canalize, facilitate, and augment what we know. There is an obvious trade-off that, out of self-justification and narrow-mindedness maybe, most teachers omit to mention: disciplined knowledge preconfigures the knowable at the very moment it raises its first questions. Ask an average historian to speak about the transhistorical or even the advantages of anachronism, and the discussion will rapidly become vitriolic. (I have tried this many times.) The *blind spots* of each discipline are most often rejected as worthless by "professionals." It is even ordinary to hear a practitioner of the discipline arguing that if such and such phenomena have no place in her own field, this is a clear sign of their inexistence. (To keep my example: everything human is historical, then ...) The opposite intuition—generally coming from outsiders—is precisely that the avoidance of particular problems tells us something about the intrinsic merits and qualities of the discipline.

Then, even within its own frame and protocols, a discipline is unable to deliver exactly or completely what it claims to supply. It is too easy to assert that everything a discipline is apt to identify is ultimately knowable and that the unknown merely names what is "not known *yet.*" Gödel's theorems posit an irreducible point of unprovability in formalized mathematics. The most discursive disciplines are also marred by a comparable "incompleteness" that is tied to the way human verbal language works. All of this is allowed by intellection, as a *defective* performance of cognition that leads to its potential *excess*. The intellective begins when this dynamic of more and less is significant.

Moreover, *nescience* is inscribed in the development of science. The course of the sciences is their curse: the more humans build, discover, and understand the real of the sciences, the more unanswered questions appear. It should be trivial to say it again today, but it is not: scientific progression is not just "progress." The solution to a problem raises another question, one that is often more difficult to answer (string theory, for instance, arises from the discrepancies between two major conceptions, quantum physics and general relativity). Then, a scientific solution is not entirely stable, and it has to be disputable. This fact is frequently overlooked, because of the persistence of a tacit positivism and the better appearance of eternal universality that mathematical objects would possess. Across the history of mathematics as a discipline, the transformation of what was the only norm into just one possible case (see the Euclidian space and differential geometries, Aristotelian logic and nonstandard logics) corresponds to the negative exercise of science. From this perspective, the sciences (and not "science") are structurally in crisis: without disputes, without the repeated creation of new zones of the unknowable, the sciences would be over. Crises condition the "advancement of science," as they do (albeit differently) with the "humanities." Because the sciences are modes of inquiry that specialize in exploring the inconstructible, they are politically perceived as inconstructible themselves—a very inadequate description. But they are not unified, equal, or faultless. The humanities are in crisis, true; so are the sciences.

I may have sketched some sort of negative epistemology by insisting on *blind spots, incompleteness,* and *nescience.* I am not pleading for neo-Pyrrhonism (saying, because nothing is absolutely guaranteed, we would better suspend our judgment) nor even deriding epistemic pretentions. I just want to consider the place of the unknown in what I know. Over the years, I took the habit of saying that thinking is *defective.* One could wish to restrict the use of this term to the second obstacle I mentioned here—the internal incapacity of the disciplinary to completely authorize itself and examine its accuracy. I will sometimes intend *defectiveness* as a more widespread property. It is one of the most striking features of human verbal language, but it also seems to belong to intellection as a whole.

14.

There are many attempts at containing defectiveness, especially on a discursive level. None of them will annul the predefined limits and structural incapacities of the disciplinary—and so far, no end is in view for knowledge (except maybe its brutal death in an outburst of globalized stupidity). Nonetheless, the old idea of the polymath could keep its charm: wouldn't it be a manageable answer to the epistemic divisions and their effects? I do respect those who want to go beside the frontiers of their fields, if they actually wish to do more than tourism in a foreign country. But while some local limitations could be attenuated by the recourse to other procedures or ideas, there will be no unification. Then, even "knowledge as such," being a cognitive and intellective act, would always extend beyond itself, that is, both failing and producing the unpredictable. Interdisciplinary research, once maximally conducted as an intense effort of confrontation between irreconcilable approaches (and not as a quest for a super-discipline or a mere collaboration), is *indiscipline*: it dwells on the gaps of, among, and between each discipline. Indiscipline makes something of the unknowable, being, then, at odds with the institution of knowledge. This is what touches the literary too, by coming afterward, by doing and undoing the previous speeches that it speaks in the production of an *oeuvre*.

For every one of us, the intensity of what we know is fractured by the unknown. "Where should we go from there?" is the question. Capturing the intellective moment is a tentative answer.

15.

The intellective space is conjectural. It is virtually accessible as soon as cognition is performed and shared. If cognition is merely operated, then what we exchange is information, with a flexible degree of success—and no additional space. If cognition is differentially performed, if this creates a relative indetermination that is motivated and reinjected into circulating thoughts, with new and unexpected effects, then our sentences and signs are being constantly, and collectively, redefined in a transiently appearing space. This is where we happen to meet, not where we are just synchronized or united by our commonness. What I described as *cabotage* is the tactic we massively use to avoid getting lost in this supplemental space of thought. We sometimes wander further and longer. The intellective (taken as a "very high order" activity) is even less durable than its cognitive operations, and it is not the *ordinary* mode of reflection.

In the last part of his career, Karl Popper strangely spoke of three "worlds," one being physical and biological *minus* the human mind and its products, another being mental and subjective, and a third one peopled by *"objective contents of thought,* especially of scientific and poetic thoughts and of works of arts."[12] Popper conceded that each world could be subdivided, and I would undoubtedly stress a movement of diffraction. If we leave aside a whole range of conceptual difficulties (concerning the status of objectivity, the role of logical deductions, and the ties between this hypothesis and the doctrine of scientific falsifiability), it seems that this "world 3," though subject to change, could mainly function as a stable repository for the products of the intellective—Popper often refers to libraries in lieu of illustrations for his suggestion.[13] This world would offer a place for the amplitude of the virtual that has to be processed by the mind (in world 2, or by cogitation). The intellective space to which we allude would be created by the need to think the excess of the *cogitatum,* as it is expressively performed. In other words, Popper's world 3 is a nondynamic and stratified approach to what we try to understand as being mobile and bifurcating. Here, doing—like Popper—as if the "objectivity" of knowledge were durably independent on physical inscription (world 1) and series of cognitive actions (world 2) may well appear to be too simple. In our perspective, we have to consider the separation of the intellective from intellection.

Time for an experiment! Let us put some human agents into fMRI machines.[14] They are listening to a story told by someone else, whose brain activity was also recorded as she spoke. The story is dull and largely dependent on the society to which the speaker and the listeners belong (an anecdote about a high school prom night in the United States). There is a level of factual comprehension that one could check with some precision. It appears that the agents with the most accurate understanding of the story exhibit more similar patterns of neural activity, among themselves and with the recorded speaker. The similarity, of course, is measured thanks to the model brain of the "average" human. It also seems that the "best listeners" have a propensity to *anticipate* slightly elements of the story—a sign both of their comprehension of narrative logic and of the banal predictability of the story they are hearing. These people have a lot in common: major biological and chemophysical constraints, the grammatical rules of the language they speak, cultural references, social conditioning, and their situation in a magnet. Their personal histories or values as bodily creatures may be without much commonality. In this experiment, the agents are involved in controlled interaction. The main concern of the scientists here is to isolate the common (visually and mathematically) in superposing the scans. The correlated operations they identify have not only "natural" causes but are also shaped by a common "culture" that influences cognition. If a factually similar story is told in a language the listeners do not understand, the brain activity among agents no longer corresponds, even remotely, to the patterns recorded in the head of the speaker. Even a partial ignorance of the social setting of the story (and the conventions of a prom night) would be consequential. This is to stress the fact that the common is not exactly coextensive to the universal. There is more. In their idiosyncratic performances of cognition, the agents consider other ideas, other feelings, other perceptions. Some of these are due to peripheral reasons that do not pertain to the story (this could be the position of the body, for instance). And a remarkable portion of additional thoughts is solicited by the story itself, owing to misunderstandings, connotations, personal memories, vagueness of the words, and so on. In both cases, the supplement would reenter the cognitive. Metacognitive processes could try to suppress some of the noise. But the most plausible explanation for the slight cerebral anticipation that is observed is that something has been added to mere cognitive decoding, something that has been integrated into cognition (such as the understanding of narrative grammar, or a rightful guess on the social consequences of the action that has just been reported in the anecdote). The participants have *shared* something that is not limited to—but is bounded by—their commonalities. Though this experiment suspends direct interaction between agents, the defective transmission of the original story (causing more and less than itself to be "understood") is shared among them: the virtual range of its differential performances describes a subspace of the intellective where minds "meet." In a live dialogue, we would certainly have much more dramatic effects, with both larger points of divergence coming from the cognitive recycling of "noisy" intellection and more reciprocal movements among agents in the direction of coelaborated attractors. In the study of a text or the viewing of a movie, the level of the agents' expertise would condition the virtual range of performances.

At a time when the cognitive sciences are rediscovering phenomenology, it is important to note that what is shared does not equal the domain of the intersubjective. Husserl's concept refers to the transcendental quality allowing the subjective experience to be recognized as commonly held (which leads to the recognition of reality beyond solipsism). Our shared space is much more transient and conditional than is the realm of intersubjectivity, and it could not be understood as perennial evidence for objectivity. Shared intellection is not coextensive to the interindividual realm either. Groups of neurons and cortical areas are already engaged in a process of exchange that potentially bypasses their constitutive operations. Plurality does not begin with the number of bodies. This is one partial implication of Plato's well-known remark about "thought" *(dianoia)* being a "voiceless dialogue of the soul with itself."[15] Here I am emphasizing the process, circulation, and exchange the prefix *dia* could conjure up. *Noēsis,* once performed, is already becoming *dianoēsis.* "If thought maintains itself it does so transversely."[16]

16.

Intellective thinking is *extraordinary, dialogic, defective,* and *affirmative. Extraordinary* does not stand for *miraculous,* even though the intense uncommonness of the intellective experience often leads to its mystical qualification. *Extraordinary* refers to something that does not belong to the regular orders of thought (encompassing mental habits, the effects of the disciplinary, or the more physical laws of cognition). *Defective* is almost a given, after what we just said, but the word needs to be stressed. The intellective could be understood as the point of defection of cognition itself. Furthermore, it is extremely transient and fragile. There is still a debate about the encryption of information at the neuronal level. It would seem difficult to have no inscription at all. But the thesis resulting in a kind of hyperspecialization of the brain (with one particular neuron coding for the actress Jennifer Aniston, for instance[17]) sounds like a very archaic proposal. A more likely hypothesis would admit both some level of material information encryption and the ad hoc emergence of ideas proceeding from multiple neurodynamic retrievals—the subsequent *recollection* (but not exact record) of those ideas being influential in the appearance of other thoughts. The intellective is also defective inasmuch as it stops at some point and has to be cognitively encoded for future uses. We have to deal here with the unexpected *effect* of what is emerging. What comes to our minds in a flash will dramatically vanish. *Dialogic* conveys the processual and i(n)tera(c)tive aspects we underlined before. By a deliberate pun, it also refers to the faculty of *language* and to the mental aptitudes the latter (in one word, *logos*) consolidates or creates. It finally designates a use of *logic* that would not be so "logical" prima facie, or the fractured logic of the "same time." Being who we are, most intellections we share are performed through verbal language, and this additional layer is absolutely not transparent. The gaps it introduces within the expressive constitution of ideas

largely participate in the distribution of nonstandard logical statements. It is here that intellective thinking is not only in defect but
also *affirmative* of its value, however evanescent, frail, incomplete,
or contradictory it may appear to be.

A particular way of conducting philosophy in the last century
was surely an attempt at embracing a certain *beyond* and reterritorializing the "Continental" on the intellective space. Over the same
period, "analytic" authors, much more classically, were doing their
best to confine themselves to the cognitive, while implementing
and incorporating, in a newer spirit, formalized tools and experimental results borrowed from the sciences. The big divide, in this
respect, was between *transformation* and *adaptation,* both camps
trying to preserve the philosophical in the end. This preservation
is sensible in the most sophisticated suggestions from "theory,"
in Deleuze and Derrida for instance, who both argue in favor of
the affirmative. Deleuze is fascinated by the differential hope of
a position without negation, whereas Derrida finds a nonoriginal
yes, which would be preliminary to any deconstructible position to
come. Both, then, intend to avoid the post hoc affirmation of the dialogical itself (the one and the others), in accordance with a structural trait of philosophy. The main distinctions among philosophical styles (but not doctrines) come from the importance granted
to what I name the intellective—and to their endurance within
the supplemental space. It is far too manifest that the dominant
practices since the Enlightenment consisted in a will to suppress
the intellective and that many authors tried to shake this dogma.
The most accomplished endeavors on this front over the last few
decades (including several key authors of "Continental" thought
and, within the "analytic tradition" itself, semi-renegades such as
Graham Priest) are worthwhile to us. But while they expose themselves to the opening of the noncognitive, I also fear that, as long as
they are bound to the rule of the philosophical, they may end their
journey a bit too soon.

17.

Should we advance the claim that "altered states of consciousness"
favor access to the intellective space through their deformation

of cognitive norms? They evidently modify the dominant logic of thought, they mark a departure from ordinary mental life and are incomplete. However, there are multiple issues with the identification of altered consciousness as the intellective.

It should be understood that the absorption of drugs, the practice of meditation, the effects of sensory deprivation, or even hunger "alter" *cognition.* These other states, obtained through different techniques, are not *less* common or *more* differential than their usual counterparts in mental life. For instance, there seem to be quasi-universal reactions to the ingestion of hallucinogens such as peyote or LSD, with marked steps that most agents will reach, involving at first geometrical figures and shapes. Or the sensation of a "presence" could be elicited through isolation and/or artificial simulation of brain areas. Such impressions have a high degree of variability, giving them content, but none of them would exclusively happen at the external gates of cognition. Some powerful and widespread mental mechanisms of control (including obedience to dominant logicality) could be softened or almost suspended in altered consciousness. This could explain why the intellective space *may* be more evidently accessible in those cases: if, and only if, the norm of a return to the cognitive is not systematically enforced, the *trip* may more easily become a *journey.* Yet, in itself, the *alteration* of the mind is just one possible genre of cogitation. The resemblances with the characteristics of the intellective space derive from a confusion between altered cognition and its differential performance. A certain kind of *madness* (the one Erasmus praised, centuries before the celebration of *folie solaire* by Michel Foucault or of schizophrenia by Deleuze and Guattari) might sometimes consist in a recurring way of getting lost on the intellective space. As long as thought is actualized, cognition remains. Then, it is just being altered *durably.* In such "madness," the extraordinary would be converted into a new order. The multitude of subnorms caused by mental "pathologies" does not correspond to an autonomy of the mind, severed from its laws, and these alternative norms are not the ultimate realization of the supplemental space of thought. They rather remind us that variability is physically bounded and that it is necessary for the intellective but not equal to it.

All in all, altered cognition is a quite *common* phenomenon. It is notoriously difficult to share with others (including the I that I am now) the experience of dreams, drugs, or delirium. This is certainly a barrier to the intellective, unless one possesses the required (and usually artistic) skills to re-create alterations in the diffraction of a now. It is even possible that some sorts of "madness" interrupt or lessen the dialogue between cortical areas. To sum up this point: the intellective *share* is not a priori fostered by mind alterations, it is just possible a posteriori, with, perhaps, a higher degree of rupture with the cognitive, owing to the release of some mental constraints and controls.

18.

If I am moved, so is my thinking. As for the intellective, it affects our thoughts in return.

19.

The intellective is a significant *tracé* on a supplemental space. A *tracé* is drawn or being drawn; it is distinctive in virtue of its own form; it is made of lines, curves, and intersections. A *tracé* is more than a *trace*; it is more strongly marked maybe, but it is also *less*, less legible than a track, and even less conceptual. Outside of the potentiality for signification, the intellective is very little, like the open-ended mistake of performed cognition. Without agents in search of something more than coded sense, this would be it.

20.

As *res cogitans,* I am in a sense *pro-duced* by my brain activity. A new project, among neuroscientists, consists in building a map of all the paths of connectivity in a brain. But what would the "human connectome" provide? Would it contain our thoughts? A first version, along the line of the "human genome project," was to give a model of the "average" connective system in a brain. Even the early promoters had to concede that, so conceived, the map would just give a certain idea of the "normal" and the "pathological" and that the mathematical and technical operation of standardization would be extremely consequential (and potentially misleading). The updated version of the project is now based on individual connectomes. The question of feasibility is decisive in all the current (and future) approaches of this kind. More than this, I want to underline that the motto "I am my connectome"[18] is confusing. While systems of connections tend to be much more stable than episodic thoughts at the individual level, it is an illusion to consider that *I* am more the paths than I am the wanderer. There is even a further fallacy, because the connectome is the map, and not the roads themselves.

In the same vein, asserting, as Catherine Malabou does, that "a brain that changes itself . . . is exactly what 'I' am"[19] is *almost* accurate, though finally deceptive. In the first half of the eighteenth century, long before the advent of modern neuroscience, La Mettrie was facing the same conceptual difficulty in his *Treatise on the Soul.* He was already equating the individual "faculties of the soul" with the "mere mechanical arrangement of the parts that form the marrow of the brain"[20] and adding that "the brain always changes state."[21] Granted, I would not expose and explore the stratification of my *I* without my brain, and the latter would not support who I am without the changes it experiences through the story of my own life or through series of cognitive episodes. But I am not *exactly* my metamorphic brain; I am at most its product, a product that coincides neither with the plasticity of neuronal matter nor with its action itself—I am a product that exceeds up to the limits of the mental.[22] Malabou amalgamates the substrate and the synaptic, both cerebral cognition with its extendedness, thereby precluding the intellective, for she seeks to posit plasticity as an arch-concept and has consequently to subsume long-term modifications, information storage or retrieval, and episodic activity under *one* rubric that would change "itself."

The intellective hypothesis allows us to recognize the constraints that are brought by the genetic, physical, and chemical structure of our brain, its synaptic functions and its plasticity, and its acquired system of connections. Then, we are able to add that *thinking* (which is precisely where *I* am now, at the moment I say it) also occurs in a noncommon and ad hoc supplemental space, where conventional road maps are of little use.

21.

Among animals, are we the only ones to sometimes sojourn in the intellective space? After all, variability in the living realm is in no way restricted to the mind, and it certainly does not "begin" with *Homo sapiens*. So, of course, there is a potentially differential performance of cognition in animals. It remains that the presentation we gave of cogitation is somehow difficult to transpose for a snail. Some animals are certainly capable of "higher-order cognition," so it invites the question of the further advent of the intellective. Before developing some "animal meditations" in the second part of this book, I will simply recast my argument in relation to several problems, as they are identified by contemporary scientific research.

First of all, if we are interested in complexity, then the number of neurons (and of possible synapses) is essential for reaching a threshold. This number separates humans from most other animals. Contrary to what is repeated over and over again, the average number of neurons in modern humans is not exactly known, but, with or without the mythical numbers of one hundred billion (plus ten or even fifty times more glial cells), the order of magnitude in *sapiens* is very different from what is found in the brains of phylogenetically close species. The type of neuron surely matters. The Von Economo neurons, for instance, which connect areas of the brain that are far apart from each other and allow additional communication, might be more common in (or even restricted to) species whose intelligence is supposedly "very high." Elephants and dolphins are endowed with these long neurons, and bonobos may be the only creatures, along with humans, to possess clusters of such cells. Other cerebral parameters of complexity might depend on the size of the brain relative to the body mass index, on gene development, on nonneuron cells, on reentry, and so on. These different lines of explanation are all being considered by contemporary science. It is fair to say that, after decades of research, the accumulated data only support basic conclusions so far.

Individual interaction (in a biotope, with conspecifics, and with other species) is another factor. It has been successfully suggested that animal societies with a sufficient level of both order and flexibility—the *fission–fusion* type, allowing frequent internal reconfiguration—needed individuals to be more proficient in unexpected situations. Cetaceans, elephants, apes, and some parrots and crows all live in such societies. More generally, it has been repeatedly established that isolation or a lack of maternal care at the developmental stage modifies cognitive efficiency.

A third factor is the manipulation of cognitive tools creating an *extended mind*.[23] Prosthetic reservoirs could both facilitate and amplify our mental tasks. Once cognition is *extended,* the extent of thought is more likely to lie beyond. We are quite sure that *writing* has been invented late in the history of *Homo sapiens* and that the event is too recent to be crucial in the genetic evolution of the species. What this extended element brings with it in terms of focus, self-discipline, and epistemic accumulation is still very palpable and is in patent relationship with a potential expansion of the intellective experience (through interaction with the deceased and the remote). "Linguistic" or "languaged" apes are other non-human examples of organisms whose cognitive skills have been enhanced and modified through the introduction of new techniques of expression.

This list of factors is not exhaustive. If we want to go up to the intellective, the capacity of *sharing* in general would be another required behavior. Sharing food (in contradistinction with distributing food, according to social hierarchies) is usually considered to be a rare behavior among animals, even though the debate is largely unresolved. Synchronized gazing among animal individuals is rather scarce and has been mostly detected among corvids, primates, and canines so far. It is often supposed to be the possible sign of an engagement and of a "shared intentionality." I can attest that the languaged bonobos reared by Sue Savage-Rumbaugh routinely use their gaze to begin an interaction or an interlocution with humans and that they also frequently share their food, more than the apes of the control group. It could even be said that, in their own recourse to human language, these apes illustrate the impact of verbal intellection on cognition. Exchanging with them is subjectively witnessing the frail addition, through speech, of the unthinkable to a more ordinary process of thinking.

In any case, we cannot speak of "the animal" in general (a point that Derrida made with "eloquence"). Some animals (including some individuals in some species) may reach a very high level of cognitive complexity. Hence, they could theoretically be brought to the doors of the intellective space. But, as far as I can *tell*, an actual journey requires verbal language, something very few non-human individuals are able to manipulate. At the very least, it requires a propensity to make things *significant*.

22.

Reality makes no sense, though the world we live in is semantic through and through. Glimpses of the no-sense are everywhere for us to catch, but they hurt our sight. Once our semantic apparatus is on, it is extremely difficult to contemplate the no-sense of reality for what it is. Even saying what I just wrote is trying to make some sense of all this. Speaking of "the absurd" is too much, conferring a particular place to what has none for us. Stoic ataraxia, Zen nihil, and toxic stupor are diverse ways to prevent our semantic functions from being disturbed by what displaces us, in favor of psychological erasure. The demiurge ultimately gives meaning to everything. The faith in complete explication through Science is another version of the same desperate gesture. There is a "Great Outdoors," as Meillassoux says in *After Finitude*; it is just that its name and meaning are coming through (and not *to*) us.

We are embedded in *semanticism*.

The worst of all is when we speak, when we perform our thoughts discursively, when we use this *turn of mind* based on the promise of some sense. We are hypnotized by the power of our own words: if "our" things (chains of sounds, weird inked shapes, holes in a sheet of paper) do mean, by which principle would external objects be asemantic? This reasoning makes sense—which is the major issue with it. Human verbal language is an extrapolation and a consolidation of semantic aptitudes. In obtaining meaning through arbitrary objects, we mimic the cognitive attribution of sense to discrete phenomena. We do not modify no-sense; we rather try to fill it. As "human normal adults," we are no longer in a mental state attuned to the unrelated. The situations we witness are potentially interconnected. They are not unique, they have similarities we may distinguish, they are linked to other moments we have stocked in our memory. We have an interest in scientific explanation. However, when a stone that has been thrown is finishing its course on the ground, we perceive more than a local illustration of the laws of gravity—would the stone be suddenly suspended in the air, this would make no sense to us, or this would be the *sign* of a miracle.

It has been successfully argued by Terrence Deacon, Ian Tattersall, and many others that, in the last hundred thousand years, or maybe even more recently, *Homo sapiens* reached a new cognitive level and newly manifested a symbolic aptitude. This theory refers to the appearance (as far as we can judge, based on the fossils and vestiges we find) of a new social organization of space in human settlements, of ornamentation and art, of more innovative material techniques. Given the time lag with anatomical changes, the symbolic faculty is seen as supported by the diffusion of the capacity for language or, *a minima,* by a drastic change in verbal fluency. But the acknowledgment of a symbol is conditioned by the possibility of making sense. This, I believe, precedes human verbal language—and is both presupposed and tremendously reinforced by fluent speech.

23.

Exposure, association, and organization are preliminary steps. The perception of a particular shape in the sky is an information. Suppose it is mentally recorded in the brain of a vervet monkey. From there, the information is comparable to other geometrical figures, and once the same kind of shape is seen and identified, it triggers a relatively regular response (e.g., anxiety, aggression). There is logical or chronological consecutiveness that is acquired by an animal through its development. Without experience, memory, referential linkage, and some hierarchy of data, the shape of a predator would be a mere percept. Here it is minimally a cue. Behaviorism saw only a conditioned response in such instances, by positing a methodical impenetrability of the mind. We may well have something like an ideation that is sustained by a system of alarm calls. Communication among vervets has been studied since the pioneering research led by Cheney and Seyfarth. These monkeys have several distinct alarm calls (one of them being associated with a flying predator) eliciting different reliable responses (such as looking in the sky rather than at one's feet) when they are emitted (even when played by a sound equipment, in the physical absence of the sender). These calls are gradually acquired. Like in Quine's colonial fable about the *gavagai,* it is difficult to establish if the types of alarm convey information about different dangerous animals, classes of predators, the localization of a threat, or the geometry of menacing shapes. I think we are authorized to assume some constant process of ideation, with a strong limitation of the incidence of variability imposed by the small number of alarm calls and the context (danger above, below, on the horizon). The additional order of association—with a referential chain of sounds—is made possible by prior mental characterization, and it stabilizes the connections between groups of percepts by ritually attaching them to a call. At the sight of a bird of prey, an adult monkey is normally alarmed, based on previous associations. The corresponding call is evaluated on a case-by-case basis and could be neglected (some individuals, such as younglings, are less "trusted" by the group). We are not facing mere conditioning; we are dealing with more than iconicity, information, or a mnemonic shortcut for a probable situation. The additional and nonconsecutive order of association allows semantic dissociation, as one step on the long path toward *semanticism,* or the proliferating attribution of mobile meanings that characterizes human thought as we know it.

24.

Our semantic milieu makes quite difficult the task of exploring the constitution of any sense among other animals. A communicative sign with referential content, eliciting a certain kind of reaction, if produced by a monkey, will be revoked as anthropocentrism: mind you, this is not a word[24] or something similar. On the other hand, the slightest gesture made by a human child is immediately interpreted as meaningful. Michael Tomasello, one of the most celebrated experts in primate cognition, finds it completely natural to provide the following gloss about an infant merely pointing at a Christmas tree: "Attend to the Christmas tree; isn't it cool?"[25] In parallel, Tomasello argues that "the ability to attribute significance to ordering patterns in learned signs"[26] shown in some chimpanzees, bonobos, and "a number of other nonprimate species, such as dolphins and parrots,"[27] could have "very little to do with communication."[28]

We are so embedded in semanticism that the "scientific" choices appear to be limited by a circular argument. Either meaning is presupposed in human infants while being refused to any other animal no matter what, or it is acknowledged to the condition of being severed from communication. Semantics is not the *apanage* of the human, no more than syntax or communication. Moreover, at least some crucial aspects of our language are transferrable to other animals (and to computers). This does not hide the fact that, based on what is currently known of animal systems of communication— and we do not know so much, to be honest—it is reasonable to assume that, on this planet, only *Homo sapiens* have had so far the ability to *invent* a tool as complex as the languages we speak. Such a probable fact does not justify seeing a "human uniqueness" in all the decisive aspects pertaining to *logos,* unless one takes the specific for the unique and differences for absolute divergences.

25.

In its constitution, human language is certainly dependent on general cognitive skills (maybe recursion, at least some logical-syntactic faculty), and, in the process, it modified (and still alters) other mental abilities. This is a speculation whose validity is made sharper by numerous scientific takes on the problem. Now, none of these expansions or "co-options" would occur without a preliminary commitment to the semantic and to the mechanisms of associative dissociations.

As soon as we stop considering an idea as a fixed point, we are ready to abandon the string of equivalence between the sign, the signified, and the concept. There is no *language of thought* (LOT), this fantastic universal "mentalese" that particular idioms would duplicate. The LOT is just one theoretical formulation of the cognitive. Thus, it is a partial view that threatens to become a self-fulfilling prophecy, or perhaps our *lot.* It also has the ambivalent advantage of escorting the scientific gesture of formalization. It is true that the terms of an equation are more constrained than even their named counterparts (the designation x over "the unknown" for instance). But it is odd to conclude that the "natural languages" could be reformed or that their gaps are of no major importance. The overestimation of syntax within language is another aspect of the same confusion: because our scientific method is more at ease with a particular aspect of a phenomenon, let us admit that this aspect is the only one or the most decisive.

Our discourses, as much as our cogitation, are submitted to physical limitations, political neutralizations, and the like. Concepts are not common; their appearance and fixation are not "free"; their frames of reference are rarely unique. Our mental processes are not endless and not a continuous flux. Words do not encapsulate rigid senses. They have domains whose exact contours vary among interlocutors, though they may be *approximately* superposed. They have different functions, and verbs or adjectives do not behave the same way, or some words are more dependent on realistic reference than others. Finding a shared semantic spectrum is the easiest way to make sense of a discourse, and all the constraints we evoked before are in use—from social norms to usage, from stylistic conventions to parlance. In the complex dynamical system of neurons, the meanings of a word correspond to zones of convergence and relative stability; they point to attractor basins for ideation. Senses are not inherent in the word; they are not independent from each other; they are not absolutely stable; they are not noiselessly replicated from one person to the other; they are more or less complicated to obtain and retain; they depend on many factors in expression and interpretation; they are not eternal. There is, however, the promise of some semantic correlation because of the common basis of cognitive processes, the perceived structures of the world of reference, and the extent of our preliminary agreements. It should be noted at this point that association and dissociation entertain each other.

Meaning could refer to the selective process across *senses.* There is a statistical probability for some senses to appear as attractors for a group of words, but meaning has to be formed on the fly. It would name the trajectory of a cogitation, channeled through the semantic potentials of a discourse. Meaning would be a dynamic engendering relative points of stability, through the analytic or "good enough" examination of an utterance. By focusing on word senses (and all the more if they are rigidified), one simply occults the way cognition is performed at several nonconcordant levels—and the ensuing incorporation of "unwanted" senses or rejected cues into the semantic fabric of understanding. No verbal definition is strict, including the ones I gave.

26.

In the intellective space, meaning is affirmed as *signification*. It is extraordinary, for it newly rearranges multiple series of word senses and prepares the singular emergence of the unheard. It is defective because the semantic attractors are as unstable as they could be. It is intensely dialogic insofar as it fissures the consistent self-identity of performed meaning and holds together ideations that are usually received as incompatible. The logical problem is not *expressed* through a sentence but is an outcome of the *post hoc cognitio* of the intellective—which may provoke a rebound outside of the cognitive range again. Nevertheless, signification is affirmed.

A well-known double-bind such as the order "Disobey!" could be heard as an impossible contradiction and as something "meaningless," that is, implying a semantic trajectory that annuls itself through its own course. If we maintain that, once it is uttered, "Disobey!" is a significant order, we need to consider the kind of dialogical rebound induced by the words. Performative contradiction could also be quoted here, as well as *adunata*. "Do I contradict myself? / Very well then I contradict myself."[29] A lack of stability in the intellective space could be found in sentences such as Gertrude Stein's "a rose is a rose is a rose,"[30] where the tautological form is at odds with the repeated emergence of semantic inflexions within "the same."

In all these examples, standard logic could be alleged to dismiss these paradoxes as deplorable artifacts of our imperfect language or as pure and void noise (a *flatus vocis*). On the "contrary," a nonclassical approach would formalize the explicit and state the implicit, then claim that the flaws of natural languages are not the only source of such problems. I am saying that discursive intellection challenges the rational reterritorializing of the cognitive.

From there, the main philosophical option is to ban the alien. Another "comprehensive" attitude goes back to possible senses and articulates them (through discourse or formal notations). Keeping something of the animadversion the paradox brought is a major difficulty. A fascinated contemplation, reenacting the impossibility, is another temptation, to the risk of errancy. The reversion to cognition (in view of a dismissal or a reappraisal) cannot conceal that another noetic mode has been induced through the consequence of discursive meaning.

27.

Language, then, is *opaque, open,* and *operative.*

28.

Language is opaque. We train each other; we train ourselves to attach affects and ideas to sounds and geometric shapes. Some of them are the envelope of words we can organize through a typified system of links, a morphosyntax. A multitude of word senses is considered, rearranged, forgotten, remembered at every one of our utterances. What we share is made of interconnected signs appearing within an idiom, a setting, a situation—signs pointing to rather indeterminate areas of the world and of our minds. An object of language is a virtual *deixis,* which is momentarily actualized through cogitation. Language is opaque because the virtual is untouched by the actual: the designation is not exhausted by its operation. As long as words stay with us, their meaning is able to be performed differently, while they are structurally identical to themselves (in their phonology, their spelling, etc.). I can "replay" her words over and over again; their senses will differ. Names, proverbs, locutions, readymade expressions, appear to us as somehow constant; a mere tone of voice may break this impression. When we are used to reading and writing with mechanical means, the sensation of word identity is certainly stronger. The verbal elements of language usually seem to stay the same, with a much higher degree of stability than our thoughts. Words invite us to perform our cogitations within a frame; they help us distinguish variations we have ignored so far. *At the same time,* the evidence of semantic approximation and variability is all the more manifest when compared to sign stability. There is a broken promise. *Words betray us.*

This *opacity* is an effect of the disposition of nonverbal intellection, an effect that is both entertained and maintained by language.

29.

Language is open. If it were the closed circle so many represent, it would *never mean anything.* Language is open to the real, to nonverbal cognition, to the intellective. Moreover, it is no more sealed than any of our subsystems of thought; it is incapable of staying within itself. A discursive critique of language is apt to refer to what lies outside of it; it is not *more* real. As long as I express my thoughts through language, they will inexorably bypass the orb of their verbal inscription, but they will never *completely* stay out of word materiality and what this entices. If, as William Burroughs said, language is a virus, we are contaminated, and there is no remedy but death. The drug of formalism is no panacea. The complementary therapy of the cognitive is just an alleviating treatment.

Were the "linguistic turn" a way of saying that there is nothing outside language, this would just be a nice try, because even language is already out of itself. Were a "realistic turn" a way of saying *verbalized* ideas are ultimately separated from language, this would be a gentle joke, because thinking is channeled through its own virtualized expression.

30.

Nevertheless, *language is operative.* A gradually invented technique whose transmission is favored genetically during human infancy, an empirical summation of semantic means, language works as a *kluge* (to recycle Gary Marcus's term). We are very far from the *perfection* Chomsky once saw as a possible "conclusion" to his work, before conveniently erecting it into a "methodological principle"[31] (the best way to reach a conclusion is to accept it in the premise, or so it seems). Certainly the perceived regularity of the external environment, the genetic endowment toward the acquisition of

language, and the "co-opted" recruitment of nonverbal regula-
tions (from brain structures to physical laws) are, as Chomsky now
advances, shaping the "phenomena that fall under what's loosely
called language."[32] Those conditions participate in the possibility
of making sense through words. One could also mention collective
norms (from common knowledge acquired through socialization
to usage), levels of proficiency and expertise, disciplinary training,
and gradual synchronization through interaction—all constraints
we evoked at some point—to get a better list of the practical reasons why we are able to delimit some *common ground* in our ex-
changes, with variable success and some approximation. *Perfection*
is out of the picture, even if we stick to syntax and avoid "what's
loosely called" meaning.

Defectiveness is not ineffectiveness, contrary to the tale of the
cognitive. There is no obligatory tragedy of ambiguity and also no
feat to celebrate with pomp and circumstance. By denying the defect *or* refuting the operation, we are barring from thought its wildest share or cutting ourselves from the matter of our lives.

31.

Language is an intellectual *tekhnē* (tool, technique, and art), articulated to our thinking and performing it. I am afraid that, for
many years to come, some scientists will continue to look for *the*
module of language in the cortex of *Homo sapiens,* or for *the* magical gene that makes us speak. This is a vain quest. Many convinc-
ing models, involving animals or computers, appear to confirm,
within the realm of the sciences, that the faculty of human verbal
language is not a mere product of genetic evolution. It has been
made possible by a range of cerebral functions and social ecological pressure. From there, it most likely coevolved with the brain
of *Homo sapiens.* The acquisition of semantic mapping through
words, as documented in dogs living with humans, or of lingual
and symbolic comprehension and (to a lesser level) production in
apes or robots is another reason for acknowledging that language
is not thoroughly contained or secreted by the human brain. The
old term of "culture" is often alleged; in my view, the category is
insufficient and often maintains "nature" as the given. That said,

if "culture" recovers the nongenetic transmissibility of characters between organisms, the concept could be roughly acceptable.

As a system distinct from the human cortex, language is self-organized according to a relative optimality of goals. A series of experiments conducted by Luc Steels and his colleagues simulated "language games" and basic communicational situations among embodied robots. In these artificial and controlled situations, phonological distinctions, syntactic features, and semantic repartitions seem to emerge "automatically." A number of notable pitfalls of verbal language could also be interpreted as requirements, as soon as we no longer assume that expression is first and foremost a theoretical and epistemic activity. If we consider, in particular, that language conveys semantic relations between social individuals who differentially perform cognition, then words have to be *vague* in meaning, lexical *granularity* is a plus, and *redundancy* is better. Within a community of speakers engaged in *praxis*, the flexible vagueness of word meaning augments the probability of *relative* mutual understanding, whereas the absence of a semantic spectrum (or a notional domain) would favor a "hit or miss" situation. It has been widely observed that human children use their smaller vocabulary in a manner that is more unusual than are most adult utterances. Sharp boundaries are not obligatorily ruled out by vagueness, but they do not constitute the only part of semantics (with all the rest being noise). Granularity is a complementary trait, where the vagueness of words is partially "covered" by the adjunction of terms, nuances, and inflexion—which proportionally multiplies the risk of semantic drift. Both granularity and flexibility also respond to the fission–fusion structure of human groups. Finally, redundancy is a way of minimizing the errors due to faulty transmissions. Redundancy, and all the "peripheral" semantics surrounding human communication, is another correction—and yet another source of additional variation.

Historical transformation encompasses the fractions into languages, as well as all the contingent developments. Political norms profoundly condition our idioms, beyond coevolution, cognitive limits, or organizational constraints. The example of the Pirahã, as far as we can tell, is quite extreme, where otherwise "universal" aptitudes mediated through language (such as narratives) seem to be prevented by social order. More generally, readymade expressions, and the framing of thought by "usage," as I explained in *The Empire of Language,* are frankly pervasive. Domination is assured through the sentences we *must* say—or censor. In this perspective, even punctual grammatical rules or stylistic uniformity could be imposed. In modern written idioms in general, the very existence of dictionaries had a crucial influence and undoubtedly encouraged the easy hypothesis of fixed senses. In our contemporary societies, the massive diffusion of speech (through telephone, television, or the Internet) is highly influential for the modifications of language use. Overall, the contacts between civilizations form a way of altering languages via the expansion of a common frame of social reference.

32.

Language comes to us as languages through speech. Attention to the discrepancies between idioms is mandatory if we want to say something of *logos.* Disparities have been rejected as unimportant by Chomskyan linguistics, mainstream cognitive sciences, and the philosophy of "ordinary language." Unsurprisingly, it has been difficult to imagine, from there, any kind of *dialogue* with the oeuvre of negative philosophy, or even with literary and philological analysis. Nonetheless, even among neo-universalists, what came to be amalgamated and known as the Sapir–Whorf hypothesis was often in the background, as a constant reminder of the differential realization of language. After dozens of experiments and sophisticated considerations about the impact of vocabulary on the perception of colors, no agreement has been reached. Given the kind of *performalism* I am advocating, I would assume that, across languages, the consolidation of ideas is not exactly similar. Asserting this is not enough. One should add that, even within one given language, senses are *not* the same (what I may now call *snow* would differ from the ideal construction you and I attach to the word in another sentence). And that, through both the commonness of experience and the semantic space we share, some meaning is co-elaborated.

This implies, for instance, that while the English *friend,* the Greek *philos,* and the French *ami* may not be exactly superposed in the range of their uses and senses, we are nevertheless in a position to think through the linguistic variations of friendship and construct a significant discourse by confronting heterogeneous conceptions. The monadic irreducibility of anthropological differentialism is an overreaction to the diffraction of the multiple. As a side note, let me briefly remark that the Sapir–Whorf conjecture cannot exclusively focus on vocabulary: whatever the reality of deep grammatical structures could be, the variations in terms of morphology and syntax plausibly exert influence on intellection.

33.

In each language, then, one will find different regimes of expression. Widespread words, terms mainly indexing regularly encountered referential objects, stereotypical expressions, parlance, create *mental routines* (relating to the cognitive phenomenon of "path dependency"). We retrieve more swiftly, and—quite often—with attenuated consciousness, the ideas that are attracted by those words and locutions. Other parts of the vocabulary are more specialized, or more constrained (by a situation, a discipline, or a mathematical form). In parallel, we do not treat all verbal utterances with equal care. We do not analyze each sentence, and we frequently opt for "good enough" readings, using a simplified heuristics that is distinct from the analytic approach. Unfamiliar word associations, contrary to banal metaphors, or ironical statements seem to require the additional help of zones in the right hemisphere (in the dominant population of left-lateralized subjects), and they take more time to be processed, fostering the appearance of "side effects." An intellective use of language could be described as being "blocked" or "impossible" for both approaches—suddenly, neither analysis nor wholistic comprehension is adequate—and, within the movement of affirmative signification, as "insightful" or "intuitive." These words abruptly *mean.* All meaning, all signification, is in act: a sound semantics of "natural languages" has to encompass a pragmatic.

34.

Human verbal language as *tekhnē* heavily structures the mental life of speakers. It both amplifies and modifies the performance of cognition. As a skill that is opaque though operative, and defective while open, language is a powerful tool for "reciprocal brain manipulation"[33] and all that ensues.

35.

Then, what we write could be a little ahead of ourselves. We may prefer to keep our thoughts as close to us as possible. I am rather trying to follow my words in the poetics of ideas.

36.

The literary is an exploration of language failures. Opacity is maximally deployed, undoing the ligatures of discursive conventions, thus opening its verbal density to the space beyond. Literature's main operation is to mean *nevertheless,* or to precipitate us into intellective thinking, as much as we can stand. The intensity of the experience has a counterimage in the propensity among readers (and sometimes authors) to stick to the superficial level of the plot or to collapse performance into stylistic ornament. Literature is the *warrant* of signification, for it is up to its oeuvres (*opera,* in Latin) to show indefatigably the possibility of unsettling the semantic reduction to fixed senses, of transgressing the deceptive rules of the cognitive. Aspects of verbal expression that are regularly seen as peripheral (such as prosody) or unwelcome (such as phonic repetitions via alliterations, rhymes, or paragrams) are heavily mobilized in poetic dictum and strongly exhibited as part of semantics. A discursive world sans the event of the literary would essentially be a timid and shameful exercise of thinking, like a monotonic series of insignificant cogitations.

A willful immersion in the literary certainly reconfigures parts of cognition, if some discursive command of the intellective is durably reached. Similarly, mastering cognitive gesture through formalization and logical associations may modify our mental skills. Blaise Pascal's two "spirits" (spirit of geometry, spirit of finesse), or *turns of mind,* speak of the long-term "habit" and impact diverse activities could have in the consolidation of what "the mind . . . does tacitly, naturally, and without art."[34] Three centuries before C. P. Snow's *Two Cultures,* Pascal's remark is strikingly stronger. It is "rare," the philosopher says, to see people harboring these two minds, but what compatibility could we hope? A simple transfer will not do. "Fine minds" are "disgusted" when they study science, Pascal notices. As for the direct application of geometry to "fine things," it is "ridiculous." *Nihil novi sub sole . . .* Could we have several turns of minds *at the same time*? Or should we suppose that, because the creative enfolding on the cognitive brings something back from the intellective, and because the journey on the additional space has to be "cognitized" (or, better, cognized), the major difficulty lies in the interruption of the literary by the scientific, and vice versa? Is it what we are looking for?

37.

It is not that "science does not think and cannot think";[35] it just has as little meaning as possible.

Because I am quoting Heidegger's famous statement, I guess I should recall that, for him, all the *disciplines* in general (even those belonging to the "humanities," such as history of art or philology)[36] are unable to think; that the Poetic *(Dichten)* seems very *close* to thinking but is not *Denken*;[37] and that even the modern philosopher (the only "thinker" of the list) is not actually thinking, or "not yet." This certainly relativizes the consequence of Heidegger's aphorism on "the sciences," and it is obvious, I hope, that I would be in agreement with none of these three positions. Heidegger is amalgamating lingual meaning with thought, while arguing in favor of oracular philosophy—and to the detriment of all the rest, namely, literature, poetry, knowledge, *scientia,* technique. We are not interested in the promotion of a mythical Speech of Being, and we identify several instances at work in thinking (including the nonverbal), which allows us to find in Heidegger's argument something it certainly does not say. I wrote that the whole specificity of cognitive enfolding lies in its belittlement of meaning: this is thinking, without any doubt, but with the most tenuous sense. In a handwritten note to his 1951 seminar, Heidegger clarified that it is science itself that does not think, and "not: the individual researchers, who possibly 'think,' but *then* not in the method of their research."[38] We are now able to propose that the disciplines (and all the more if they are formal and rationalistic) canalize thinking in such a way that the intellective—which is required by any strong epistemic creation—has to be abandoned or occulted to the profit of the cognitive. This implies in particular that what scientists *think* has to be unclothed of its signification as it is being prepared for transmission to the other members of the community. "Individual researchers" may have a wide array of tactics, helping them to still relate to the beyond of their noetic creations. The "method" I am using here is a manner of verbally exceeding the scientific.

38.

I have called several times for dialogic and situated there a possible aftermath for intellective contradictions. The status of logic is a topic that has been saturated by the professional practitioners of the "analytic" doctrine, who often consider themselves to be its *owners*. This is not to dismiss beforehand the validity of the formal notations taught and introduced by these philosophers, not at all. I am saying that the proprietary attitude has become a powerful way to eschew most significant questions, to the profit of technicalities. So before playing with symbols, we have a few considerations to make.

Logic is a noetic activity of "automatic" derivation between elements. Mark my words if you want, though this is less a definition than a qualification. Logical reasoning is accessible to nonhuman cognitive agents, including computers and many animal species. Despite some political resistance to the results, it is plausible that other animals, including apes or crows, are apt to reason logically. Ruling out all possibility for inference in the animal mind to the benefit of mere associations is, in my view, not tenable. One could speculate that logic is an adaptation, based on the extrapolation of geometric regularities observed in the environment. In particular, the impossibility for two perceptible and dense bodies to occupy the same position of space-time without being *the same* is a fact of our common and ordinary experience of the physical world. Logic is a cognitive shorthand that does not begin with mankind. It procures an advantage in the understanding of the surroundings, but it heavily depends on what has been abstracted. Furthermore, the discrepancy between statistical regularities and what actually occurs yields another difficulty: our logic interprets relations by correcting the unusual. This makes it a poor tool for the unpredictable, for instance.

However, logical linkage is so consubstantial to cognition as we know it that I would doubt one could ever humanly *think* without any logic. "Without any," for there may be several kinds of logic, or different uses and intensities—and not just the one seemingly structuring these demonstrative sentences. For instance, was speaking of something both here and not here, or of a cat both living and dead, absolutely unheard of before Erwin Schrödinger proposed his famous story to illustrate entanglement in quantum physics? Of course not. The (relative) novelty was that modern science would appear to adhere to such deviant considerations. I bet that all the peoples on earth *(Homo symbolicus)* have at one point considered ubiquity or the existence of the living-dead. Such ideas may be rejected as fantasies (according to some high level of material improbability), but they could also be articulated in a relatively necessary way in legends and religious doctrines. The nontrivial persistence of some contradictions structures Indian Jainism or Hui Shi's "school of names" and many other discursive reflections in Asia. In the European Middle Ages, a comparable architecture was integrated in otherwise standard deductive reasoning by Thomas Aquinas. And generally, the more we speak, the more we should become attuned to our uttered contradictions and contrarieties, as experienced with the literary. Logical linkage is based on repeated observations in a certain context. Now, the milieu by default for human cognition is not only the world of "nature" or inert bodies—it is also cogitation as such, and discursive argumentation. If we believe that an evolutionary advantage induces logicality, then having multiple logical lines, depending on their planes of operation, could be another asset.

39.

From the unpredictable, we can learn the logical inevitability of the defects of logic—but not terribly more, I'm afraid. Affects are often regarded as unexpected, or beliefs as "irrational." Nonetheless, they are not contingent, and they do not appear to us as such. The fact that thinking is affected gives way to the assumption of another order of thoughts, as soon as we do not locate the extraordinary in a sentimental perturbation of perfect logicality. We also dream a lot and sometimes wander with a lack of focus or consecutiveness. But abstraction proceeds from an ability to record, whereas dreams, raptures, or mental fugues are characterized by a loss of memory.

In contemplating the recurring defection of cognition through its performance, we can intuit the virtual existence of several logical lines. I would add that we can habituate ourselves to *dialogic*— which implies another automatic derivation than the one we need most for our survival in the external environment.

Reflecting on the statistical constitution of observable matter, and with the guidance of some innate mechanism for derivative acquisition, we form a "logical" aptitude. Because this faculty primarily depends on the mutually exclusive positions in space and time of different objects, it is filled with standard and bivalent reasoning. Once logicality has been acquired, several modes of linkage could be acceptable, if their consolidation is possible, especially via semiformal evocations through discourse. By coming later and being less deeply inscribed, they may pass for unusual, special, magical, or fantastic. But in fact, we are quite used to them, for the dynamic development of verbal meaning constantly leads us to contemplate what René Thom called the "confusion of actants."[39] Repeated incursions into the shared space of intellection have other long-standing effects than the exploration of the forest. The multiplication of logical lines may form the best option vis-à-vis the deviation of intellection, even though they are at odds with their own use in the predictability of automatic derivation.

40.

The logic (or logics) I am essaying to refer to is *implied* in cognition. It is not *applied* to it. When we think, we are logical animals, without thinking about it. Our cogitations mobilize an *informal* logicality, which we do not need to expose to make it work. The moment we express logics as sets of rules, we try to make them more logical than they are. They are no longer acquired safeguards shaping cognitive episodes; they are abstracted from the implied abstraction of regularity. They do not qualify statistical phenomena in a system of the observable world (be it "natural" or "cultural"); they are transcendental laws, independent from the place and time. Refining logic begins *semiformally*, with discursive explanation of the structures of reasoning: this goes from parts of a discussion when someone is telling another "but you are contradicting yourself by

saying that . . ." to Aristotle's chapter gamma of the treatise traditionally titled *Metaphysics*. *Formalized logic* tries to complete this
movement through conventional notations. One could say an implied *formal* logic exists in each mathematical formula, prohibiting
us to write $1 + 2 = 3 = 1 - 2$, or $(d\|f) \Leftrightarrow (d\perp f)$, for instance, if we
use conventional symbols for numerals, arithmetical operations,
geometrical entities, and relations.

In speaking of *formalized logic*, I want first to insist on discontinuities. In modern propositional logic, once fixed definitions and
rules have been given, there is little to no leeway in calculation. In
animal cogitation, logicality is variably binding. Then, the performance of cognition could simply suspend its own regularity. What
is demonstrated through the recourse to formalized logic could
give an idea of what happens in the production of our thoughts, but
it does *not* mirror human thinking. This "model," including simulation through computers as long as they are invariable and without
bodies, is so corrective and prescriptive that it is barely isomorphic
with what it claims to replicate. What it delivers is a simplified and
explanatory route toward the cognitive (and not cognition). Apart
from its value in mathematics, formalized logic grants the measure
of an operation and a propaganda tool against the excessive performance of thought. However, there are two interesting uses of
propositional logic. One is negative: in comparing the trajectory of
the intellective with its formalized rendition, we may better view
what is captured and what is still inassimilable by the scientific.
What formalization does render, however, suggests there is more
soundness of the informal than some scholars would argue. This
is the second use, through which it could become less tenable to
dismiss, say, poetry for its total lack of logic—or to celebrate art as
an act of arbitrary imagination and discard its cognitive condition.

Logic is discontinuous. As an animal operation, it does not exist
in the same way software guides computation: it is unexpressed,
often insufficiently controlled, and variably implied. As a more explicit system, it immediately bifurcates from what is its actualization on the fly, through correction and simplification. Moreover,
we do not have one logic but several—this is the case within the
realm of formalization, too, with all sorts of gaps and mutual incompatibilities. Depending on the context and the moment, we

simply opt for this and that logic over others that could be more favored by other cultures. Finally, any human *logica universalis* would
be incomplete.

41.

The mathematicians who founded modern propositional logic
intended to break away from what they perceived as the terrible
limitations of human language. Frege wanted to *write ideas* directly
(he called his system a *Begriffschrift*), without the impenetrable
mediation of common words.[40] Alfred Tarski was convinced that
"colloquial language" is so "inconsistent"[41] that "in that language it
is impossible to define a notion of truth or even to use this notion
in a consistent manner and in agreement with the laws of logic."[42]
For both authors, the "semantic antinomies" of human language
were its essential deficiency. Then, quite logically of course, the
philosophical heirs of such scholars came to believe that "every-
day language" was susceptible to be explained through formalized
logic, as long as what Carnap called "semiasology"[43] was out of the
picture ... The first "cognitive revolution" pretended to reach a logical level through language, to the exclusion of meaning. "Formal
semantics" came in a bit later on, applying formalization to "meaning," as long as it is "abstracted away from those aspects that are derived from the intention of speakers, their psychological states and
the socio-cultural aspects of the context in which their utterances
is made."[44] It is no surprise that formal semantics is still wondering
if bachelors have wives.

If language is an empirical summation, and considered in relationship with cogitation, then, yes, we find logic in many of its
dimensions. Grammar is certainly an attempt to regulate and order expression and was doomed to demonstrate some logicality,
given the role the latter plays in cognition (among humans and
other animals). Same remark for discursive reasoning. In short,
language does not kill logic; it does not invent it; it may aggregate
some of its aspects, through morphological and syntactical regu-
lation; it repeatedly reinscribes logicality in the discursive quest
for pertinence; it fosters the move from the informal to levels of
formalization.

The logical apparatus even prepares, or perhaps authorizes, semantic linkage, because it turns relative regularities into positional identities, articulates them through derivation, thus predetermining an acceptable range. Were there no determinable identity of identity, we could still not deny that some level of confidence in the logical same still operates in language. This confidence, minimally perceptible in the internal cohesion of the sign or in phonological rules, also constraining the use of some words in context, and more explicitly deployed through grammar, is at odds with everything that is supposedly senseless but is nevertheless computed into meaning or signification, and with the intellective aftermath of what has just been said *differently*. Expression urges us to reconsider the logic of our thoughts. In the cognitive-becoming of the beyond, we can opt for a reversion to bivalent and noncontradictory modes of rationality, or we can reconstruct another logic. Most of the, time, we are just somewhere in between, and we may want to stay blind to the underlying abyss.

42.

My logic is sometimes insufficient or it stops too soon. And sometimes, what is an internal lack is also excessive, giving the elements of a mode of thinking beyond the code that informed the operation of cognition. There, I assure you that colorless green ideas may well sleep on the grass.

43.

There is a discordance between informal and formalized logics, independently of the *line* that is being chosen (bivalent, trivalent, . . . ; consistent, paraconsistent, . . .). If there were none, and if "logic were logic," one and the same, shouldn't we have less trouble acquiring its formalized description? Arguing that the pure rationality of thought is simply obscured by our sentences (but rationally retrievable) is a bit short. Let us say that we have to deal with negation. This is p and $\neg p$, clear cut. What is now $\neg p$, when p is "this man is visible," exactly? Is it "this man is not visible," "this man is non-visible," "this man is invisible," without mentioning modalities ("not visible at all," "no longer visible," "half-invisible," "almost visible," etc.)? Of course, there is a classical answer to this question, but when I select one sentence as "equivalent" to $\neg p$, what is the status of the others? Are they unrelated to our problem, really? A "fuzzy logic" seems to be more able to formalize an amplitude of truth, with a whole possible range of values going from 0 to 1 as *extrema*.

Is it that, in our conversations, negation is "fuzzy," or is it that formalized fuzziness is a tentative repair for the structural inabilities shown by standard propositional logic? Or both? Unless informal logic is more inconsequentially capable of changing lines, giving us the impression of approximation and fragmentariness that decimal values of truth try to capture. I doubt I could completely respond to that, either discursively, or formally, or both—if the difficulty is due to the gap itself. No demonstration could save us here, when the very *qualification* of an informal logic $\mathcal{L}_i$ is de facto and already preformalizing $\mathcal{L}_i$ into a subpar image of the formalized logic $\mathcal{L}_f$. The truth of $\mathcal{L}_i$ is not determinable by $\mathcal{L}_f$ at a level $n + 1$. An extrapolation of Tarki's method would simply be inaccurate here and would precisely insert us into circularity, instead of adopting the ziggurat-like architecture of his proof.

44.

Through exercise, we can try to "perfect" our logic. But each time we think and experience a significant alteration, we understand (or believe) we are dealing with more than error. The logicality of our cognition may consecutively be modified: more lines emerge, and more than one is broken or crossing another path.

45.

Contradictions may arise from an *imperfect* assemblage of haphazardly gathered ideas, doctrines, and conceptions. Pragmatically, a lot of human institutions are self-contradictory—and far from the Marxist predictions, they are quite able to survive "illogically," with proper care. Contradictions may be a mark of *incompleteness,* the region where one should not go, to preserve the power of the cognitive. Then, contradictions may be significant in the intellective space. They come with the defectiveness of thought and are themselves defective too. An enduring misconception argues that it is impossible to think in contradiction. Well, that is still what we regularly do, and pretend to forget.

The principle of noncontradiction, Aristotle says, is a fundamental *axiom.* Accordingly, the famous text where the Philosopher formulates it is everything but a demonstration. For Aristotle, there is no reality and no thought without this principle. Then, his strategy will consist in ridiculing the opposite opinion and asserting that, in such contradictory conditions, everything could be said (or reflected, or done). Łukasiewicz, as he was building what became the Polish school of formalized logic in the 1920s, explained that Aristotle's inference was arbitrary. Having both $a \land \neg a$ does not have to entail any β. In the last few decades, these remarks have been largely developed by authors such as Newton Da Costa, Graham Priest, and their collaborators. What interests me at this point is the third evocation of the rule of noncontradiction in the *Metaphysics.* Beginning with a law that applies to nature, Aristotle insists on its absolute necessity for thought as well: without it, no reflection of any kind would be possible. The text says, "It is impossible for anyone to conceive that the same thing both is and is not."[45] There is a reference here, whose source is immediately given by Aristotle: "it is impossible for anyone to conceive that the same thing both is and is not, as, according to some, Heraclitus says. For it is not necessary to also conceive what one says."[46] The second-hand quote of Heraclitus ("according to some") could allude to

Aristotle's own absence of certitude on a text that was reputed to be "obscure" as soon it was published; it could be a simple mention of a lack of direct knowledge of an opinion generally attributed to the thinker of Ephesus. These words also insist on the fact that all these contradictory opinions are *uttered*—and not automatically *conceived*. The Aristotelian position is a plea for the cognitive: one should not always try to think what one says. This is a much richer and generous assumption than the final advice the *Tractatus logico-philosophicus* delivers.[47] Yet, the remark of the *Metaphysics* would force us to consider that everything discursive bypassing the cognitive is just "noise." In that case, why is this negligible babble the pivotal problem here? Why do we even have to battle against it?

In the perspective I defend, contradictions may bring us to a dialogic space. As for the famous "principle," besides the fact that it is defined so diversely, one could wonder if it is *one* and an *axiom*, I believe it to be incompletely respected in our own speech. We regularly pass through contradictions; the way we try to *do* with them is above all what matters. "Another" Wittgenstein, in the late 1930s, alluded to the importance of seeing "contradiction in a wholly different light" and to prefer what he called an "anthropological" viewpoint over the one of the "mathematician."[48] We must not convert ourselves to a kind of omnipotent regime of contradictions, which would dissolve the shock it represents for the dominant part of our logicality—but we should refine the use we already have of dialogic signification.

46.

It is difficult to distinguish what Heraclitus's doctrine was about. His most famous quote *(panta rhei)* is quite certainly apocryphal. His teaching is amalgamated with the positions of Cratylus by Plato, who may well distort the texts for his argument's sake. And because even Aristotle, whose nickname was "the reader," proceeds as if he had little or no direct access to the books of Heraclitus . . . In the fragments we have of the "Obscure" philosopher, we read multiple antitheses and oxymora. An aphorism seems to provide some advice on the right way to read the corpus; it states, "Those who have not listened to me but to the saying say wisely and likewise that the one is all things."[49] There is a disjunction between "me" and "the saying"—*logos*, here, is the act and ability of speaking, the fact of saying something (the verb *legein* is used just afterward), and also the apophthegm itself. This gap between the speaker and the spoken was a part of Aristotle's argument (nobody has to believe what he says). Heraclitus was already proposing to give precedence to words: the *I* is transported by its own discourse and may say more than what he says. The art of listening discovers, through the diversity of spoken things, a hidden unity, that precisely sustains the (fact of) saying. This underlying unity allows juxtapositions such as "immortals mortals mortals immortals,"[50] where, as often with Heraclitus, the ellipsis of the verb reinforces the conflation of antonyms. In another self-reflexive commentary, the author advances: "they do not get how the saying of the different is like itself: a reversible tie, as with a bow or a lyre."[51] Human speech ties the differences "at once," and it defies any principle of noncontradiction. This art of binding is compared to a chord—attached to the arc of the bow or the U of a lyre—that is able to oscillate and describes a curvilinear movement in one direction and the other. Somewhere else, Heraclitus adds, "In common: origin and end on the course of the circle."[52] According to this *logos*, the whole experience of the human (as thinking and speaking agent, and as immortal mortal) becomes circular. Similarly, a possibly metaphorical definition gives "a path: upward downward, one and the same."[53]

In what is extant from Heraclitus's *Physics*, one finds three main poles: *phusis*, both nature and reality, which "hides itself";[54] the *I* that "says";[55] and the god, especially the enigmatic and oblique Apollo, that is, "the master whose is the oracle in Delphi, [who] neither says nor hides but makes a sign."[56] The divine and oracular activity is described as *sēmainein*: "making a sign" or "signifying [an order]" are the most common "senses" in the Greek of the sixth century B.C., though "signifying" is also very close. Facing a reality that has no meaning, and a way of saying everything in a circle, the god makes a sign toward signification. Apollo opens up the crypt of *phusis* and cuts the total reversibility of human speech. In a more secular way, signification is like a helix, or a combination of geometric *rotation* and *translation,* that is produced through the action of discursive thinking and because of the defective circularity of language or its both intrinsic and performative openness.

For Heraclitus, as in the metaphysical tradition that follows his oeuvre, the divine is the condition of possibility for a saying that would break free from the complete circularity of language and reach *phusis.* The obscurity of style strategically echoes the event of signification in the Delphic oracle. Contradictions, then, are no longer the tokens of a closed verbal system, they are the *signs* that signification is possible. This is a "divine" way of going "beyond the limits of thought," or at least, trying to do so. For us, the signs of contra-dictions are both the marks of the real (away from any *present* reality) and the potentiality of affirmative thinking through the defection of rationality.

47.

Straightforward colonialism admitted that indigenous minds were full of "paradoxes," aporias, and contradictions. Indeed, human civilizations hold innumerable contradictions, often called magic, religion, wisdom, history, poetry, society, and so on. The zone that calls itself "the Western world" was and is no exception. Historically, because of the influence of (post)Greek philosophy and, above all, modern science, a particular insistence has been made, among meta-European civilizations, on the requirement of the cognitive. This move might be expressed as method, rationality, Enlightenment, progress, and so on. I have shown elsewhere that modern colonialism—in addition to its economical, social, and religious roles—was also an attempt to distribute the idea that "the West" was "rational" (which it was not, according to its own criteria), whereas subjugated peoples were not logical.

The presence of contradictions has been heavily emphasized by colonialists. In the reform of imperialism, giving more way to "tolerance" toward differences, a *paradoxical* celebration of other modes of thoughts could appear. This is the premise of mystifying conceptions about "primitive thought." In the first half of the twentieth century, Lucien Lévy-Bruhl argued repeatedly that the mind of "the primitives" was ruled by a "principle of participation,"[57] according to which "beings, objects, and phenomena are able to be—in a way that is incomprehensible to us but quite natural to the eye of the primitives—both themselves and something other than themselves."[58] It is to Lévy-Bruhl's credit to have finally understood, in his last notebooks, that both *participation* and *noncontradiction* were observed by the primitives as well as among "us." Alas, a multitude of scholars continue to tell the tale of human societies with no "Western" rules of thought. In parallel, many reformed anthropologists make the case that, because we do not think in contradiction, neither do other human groups. Such efforts are vastly misguided. The more recent—and "Anglophone"— debate about "Zande logic" (based, however, on older observations by Evans-Pritchard) is perpetuating many of these preconceptions. The question should not be, Do the Azande have another module of logic than "ours," for which contradiction would not be the end of it all? Or are they "like us"? As Newton Da Costa and collaborators write, if one asks "is there a Zande logic?" a good answer is "yes, there are many";[59] and, I insist, this answer would be similar for "us." The anthropological and social reality of logicality resides in the kind of particular emphasis that is given to certain logical lines. As much as a bivalent frame is presumably universal across many animal species (including ours), a wider entry into the dialogic is accessible through semanticism and complex intellection. Logical variability is enacted across individuals, species, civilizations, languages, and situations.

48.

Incompatibilities go with contradictions; or not. Contradictions go with incompatibilities; or not. Contra-*dictions* are mentally found and discursively expressed. Incompatibilities are just everywhere.

49.

"The incompossible is irreducible to the contradictory, and the compossible to the identical," Deleuze wrote in *Difference and Repetition*.[60] It could well be that contradictions form a blocked understanding of the differential process of "vice-diction."[61] This would be close to the changes I see in the formalization of the informal. But trying to keep vice-diction *immune* from any contradiction sounds like the different return of a well-known philosophical tactic—and, in fact, bipolar concepts continue to inform Deleuze's writings, up to "the two halves of difference,"[62] or the two kinds of repetition,[63] and so on. The banishment of contradiction is what we should vote against, but we also want to keep some "wilderness" here.

In *The Logic of Sense,* Deleuze adopts a different approach and suddenly celebrates the "contradictory objects"[64] (such as Meinong's square circle or the paradox of a *perpetuum mobile*), belonging to "the impossible" and having an ideal "extra-being."[65] But Deleuze also maintains that "the principle of contradiction [*sic*] applies to the possible and the real"[66] and surreptitiously ratifies both the bivalent stricture of a philosophy of "things" and the inscription of the logical within the incompatibilities of *phusis.*

50.

"Here is a tall small mouse" could be said of a large rodent for its (little) size. Or else these words are an imperfect arrangement, because *tall* and *small* are not compatible attributes. Unless something more is tentatively conveyed, and we are facing a contradiction—an agrammatical absurdity, or an enigma. But we are no longer children, are we? So we are not going to spend hours wondering. "The tall mouse is a small mouse" may arrest us more. The internal rhyme is more audible and is included in a more alliterative and rhythmic expression, as if *tall* and *small* were mirror words and *small* like a partial paragram of "tall mouse." The logical propositions do not differ that much, however. "Here is a tall small mouse" is simply representable by $t \wedge \neg t$, which could serve for the other formulation as well: the conflation of the two adjectives corresponds to the assertion of two contradictory propositions, independently of the original word order. Although we are supposedly dealing with *the same* contradiction, the second sentence seems more difficult to discard immediately and retains an oddness that is neither reduced formally nor purely ornamental.

Fortunately for our thoughts, they might be provoked by more sophisticated puzzles than this one.

51.

Literature is a significant attempt at thinking through contradictions, and not only against or after them. The role of contradictions infused a whole branch of (post)romantic literary theory, exemplified by an author such as Maurice Blanchot. This focus is historically situated and could be seen as a response to the promotion of the cognitive in the dominant (and Kantian) part of the Enlightenment. It would be pointless to claim that literature *became* an affair of contradictions after 1789 and that it was thoroughly different (or even nonexistent) before that time. The critical emphasis changed, it did not invent the property: a large part of the debate was in place in the accusation of poetic lies that Xenophanes formulated, decades before Plato.

An impressive instance of significant (and performative) contradiction could be located in a fifth century B.C. play by Euripides, where Heracles speaks to his friend Theseus. At one point of the interlocution, Heracles says the following sentences: "I do not think, have never believed, and will never admit that the gods have illicit love affairs or bind each other with chains, or that one is master of another; for what a god needs, inasmuch as he is a god indeed, is nothing—and the rest is the unfortunate words of the poets."[67] Since Antiquity, these remarks have been perceived as an imitation of Xenophanes. In the fragments we have from the philosopher's *Silloi* (or *Mockeries*), we find the idea that the representation of the gods, as it is being delivered by poets, is fallacious: "to the gods, Homer and Hesiod attributed everything that, among humans, is outrageous and blamable—theft, adultery, and deceit."[68] The general attribution of human feelings to divine entities is an error (and, say, a ridiculous contradiction in terms): "but the mortals believe that the gods had a birth, and that they have the same clothes, the same voice, the same body as they do."[69] On the basis of such quotations, Karl Popper, Theodor Adorno, and Paul Feyerabend saw Xenophanes as the epitome of the Enlightened thinker—the three scholars drawing vastly divergent conclusions from this common assessment.[70]

Is a tragedy, inspired by the same legends Homer and Hesiod used, the best place for the critique administered by Heracles?

Some commentators have argued that, here, Euripides is attacking
the old poetry and is making the case for the supremacy of tragedy. The difficulty is larger, however, because Heracles (etymologically "Hera's glory") expresses himself after having murdered his
own children and wife in a moment of madness provoked by Hera.
The goddess was seeking revenge because of the adultery that led
to Heracles's birth. In other words, Heracles on stage is the ideal
example of the validity of the myths he is discursively censuring.
Many philologists, because they belong to an ordered discipline,
"logically" admitted that either the text was interpolated or that
it was absurd, Euripides "forgetting" the setting and taking the
weird opportunity of this *stasimon* to express some unrelated considerations about the gods. To me, we are facing something else: a
critique in act of the philosophical, ushering us to significant contradictions. These words cannot be said by Heracles; but he utters
them; so, how can we think them?

52.

Here is a stanza from a poem by Charles Baudelaire:

> I am the wound and the dagger!
> I am the blow and the cheek!
> I am the members and the wheel,
> Both the victim and the executioner![71]

There are innumerable ways of not reading what this text is saying,
and the choices made by a few English translators of the original
French show that a logical understanding is at stake. It is very clear
that some difficulty has been perceived, leading to various endeavors to attenuate the cognitive "scandal" of Baudelaire's piece.
One translator adds, in the third line of the stanza, an adversative
"yet" that is nowhere to be found in French ("The limb, and yet
the wheel"[72]). Another marks typographically a disconnect with
an em-dash ("I am the wound—I am the knife"[73]). Several erase
the insistence on contradiction introduced by the original text at
the end of the stanza: while there is a repetition of *et* (and) before

"the victim" and "the executioner," which exactly parallels "both" in English, a unique "and" is favored.[74]

Baudelaire's text is quite famous and borrows its title from a Greek term used by the Latin author Terence as a title for one of his comedies (*heautontimoroumenos*, or the one that punishes himself). Throughout Baudelaire's poem, an address from an *I* to an enigmatic *you* installs an intricate and reversible relation of violence and love. After the stanza I quoted, in the last lines of the text, the *I* defines himself in the terms "I'm the vampire of my own heart," situating the result of a movement of reciprocal possession (beginning earlier with "She's in my voice, the termagant / . . . I am the sinister mirror / In which the vixen looks"[75]).

The four verses we are reading illustrate the extraordinary and impossible movement of metamorphosis and enfolding. Their informal logic is disturbing. We might want to consider that each line simply expresses the conjunction of two different logical propositions (e.g., *"I am the wound"* and *"I am the dagger,"* or $p \wedge q$). Those pairs of propositions, however, are not randomly associated, and we just saw signs of cognitive resistance in the translators' choices. While the negation of *"I am the wound"* is not restricted to *"I am the dagger"* ($\neg p \neq q$), the paradoxical—or maybe impossible— connection of the two assertions is still perceptible.

I am also under the impression that the verse "Both the victim and the executioner!" expresses the outcome (is it the *truth*?) of a dynamic that began before the stanza (especially with the first "I am" of "I am the sinister mirror . . .") and ends with the poem (with "the utter derelicts, / Condemned to eternal laughter, / And who can no longer smile"[76]). This impression, of course, is supported by supposedly extralogical elements we would have no means to formalize: the anaphoric "I am" ; the repetition of "and" ; the scansion granted by exclamation points; the rimes, the meter, the echoing rhythm (for lines 2 and 3), the alliteration of [u],[77] all adding another parallel and/or cumulative functioning; the alternating genders for the attributes of what *I* am (dagger, blow, member, and executioner are masculine in French; wound, cheek, wheel, victim are feminine), and so on. Thinking within language, that is, within a language, also shapes the logic of expression. In other terms, we

might consider that the text is showing the way leading from contrarieties to contradiction—or that it is a subjective variation on
the noetic experience of the impossible.

In a standard propositional system, an assertion of the type
$a \wedge \neg a$ ("underpinning" or "consolidating" the line "Both the
victim and the executioner") is trivial and would properly entail
*any*thing. This interpretation sounds to be incredibly restrictive
for the reading of *The Flowers of Evil*. "Yet," it corresponds to some
major approaches to the literary (or to the poetic, or to Baudelaire's
oeuvre, etc.). The accumulated contradictions could "reveal" that
these statements are false, because they are inconceivable logically.
They would be *lies,* or absurdities intentionally presented as truth
and/or fiction, or fabricated objects to be admired in virtue of
their unreality. They could be seen as a mark of *mental instability*,
or *madness* (the biographical *doxa* would remind us of Baudelaire's
syphilitic stupor in his very last years). They might be said to be
"undecidable" within the system of the poem. They could finally
illustrate the purely ornamental dimension of a work of art, whose
meaning would be indefinitely postponed, leaving us the enjoyment of the aesthetic in the meantime. All such interpretive gestures rely on a conventional acceptance of cognitive logicality, even
in the absence of actual formulae in their gloss. All of them attempt
to reduce the intellective challenge of the poem and, in this, ratify
the rationalistic program of cognitive enclosure.

53.

Claiming that *no* logical line is implied in poetry, or that formalization captures *nothing* of what is deployed in language, is very
dubious to me. It remains that, even at their best, algorithmic approaches and formalized logics would *freeze* signification. Unfortunately, at the semiformal discursive level of literary criticism,
humanistic scholars themselves often oscillate between the blunt
correction of inconsistencies (from the translators' simplifica-
tions and philological "emendations" to the positive explanation
knowing only that *A causes B*) and the suspended celebration of
the asemantic through semantic modes. It would be simplistic to
dismiss the formalizations of a poem, while maintaining in the

discourse of literary criticism a semiformal reasoning that, step by step, is just roughly (and "weakly") analogous. So let us imagine here a new brand of experimental criticism, where the attention to poetic thinking would both encompass an interest in "logistic" and entertain doubts about its "curative properties" for (extra)ordinary language.

We are going back to the same stanza of "The Heautonti-moroumenos" and choose to begin with the movement between vice-diction and contradiction. The cognitive difficulty we have with "I am the wound and the dagger!" could be transcribed as the conjunction of the propositions p and q with their asserted incompatibility:

$$(p \wedge q) \wedge (p|q).^{78}$$

This rapidly borders on contradiction. One usually admits that such incompatibility corresponds to the negation of a conjunction (in electronics, this is the NOT AND or NAND operation).[79] So we would end up with

$$(p \wedge q) \wedge \neg(p \wedge q).^{80}$$

Then, the fourth line of the stanza is a stronger case than the other verses. In the world of this poem, there is *no bystander*: *"I am the victim"* and *"I am the executioner"* (say, respectively, p' and q') negate each other but are still asserted together. With $\neg p' = q'$, we obtain for the last line of the stanza

$$(p' \wedge \neg p') \wedge \neg(p' \wedge \neg p').^{81}$$

Once again, the incompatibilities of the text prepared the expression of contradiction, or something that is "logically" forbidden, excluded, or meaningless. Now, were we to accept transconsistent sets, our equations would still have tremendous difficulty grasping the intellective promise, as delivered by Baudelaire's text.

We could first use Newton Da Costa's C_1 and introduce two types of logical negation: a strong one that is always constrained by the principle of noncontradiction (conventionally, this would be noted $\neg^*$) and a weak one having no such limits (or $\neg$).[82] With such a device, the logic of the stanza would be distinguished from an accumulation of more "explosive" contradictions[83] and would no longer be overly problematic. This weak negation might be more (just *more*) apt to underline the fragmentations of negation in natural languages. The intellective experience of contradiction is always "for now," it is not eternalized as formalism is. This solution helps us advance a formalizable *coherence* of the intellective expression that is nonetheless *nonconsistent*. It also stresses the partial, eventual, and local aspects of (literary) statements. Conversely, the double system of negation tends to erode contradiction, or even to normalize it, as has been often reproached to Da Costa's efforts. It also has little to say about the stronger case of contradiction we may have found in the last line of the stanza.[84]

Another method would be to tolerate a kernel of contradiction, while attributing to it a truth-value being neither false nor true, nor in between. Its truth would neither be a basic 0 or 1, nor a fuzzy value lying inside of the]0;1[interval. Then, within a bivalent scope, there would be nothing more to compute. But the possibility of a beyond could be inscribed, to the condition of *forcing* bivalence (i.e., of changing scope) and including, as Łukasiewicz once suggested, the possibility for a proposition to be something else than 0 or 1, such as 2. We would retrieve through this notation the conception of a *higher* poetical truth, transcending the usual limits of true and false and arising from the lacks of our rational and binary limitations. This, I believe, is quite close to where a "dialetheic" understanding would lead us: in Graham Priest's *LP* logic, a poetic enigma could compare to the liar's paradox and proffer true contradictions.

In a recent proposal, Jean-Louis Dessalles recast Leon Festinger's "cognitive dissonance" as *cognitive conflict* or an argumentative—and less static—incompatibility between different evaluations of a state of affairs (e.g., one or several lines of Baudelaire's poem). The attributed numeric values are positive or negative—zero corresponding to a null evaluation—depending on variations of acceptability (based on belief and/or desire), which dispenses us from the otherwise pervasive reference to *truth*. In this case, we could have a tension between at least two possible evaluations, for example, between a *positive* face-value in the reader (I adhere to what the *I* says in the poem) and a *negative* feel of an *adunaton* (how could this be?). Then, whatever the "actual" values could be, their combination (understood as an arithmetic product) would be negative too, which is, by definition, the sign of cognitive conflict.[85] The incompatibility becomes, in Dessalles's system, the point of departure for a reasoning using *abduction* (causal identification) and *negation* as tools for finding a solution. A skillful reading would just be blocked in a succession of conflicts, but, instead of merely "abandoning" (like the "computer" would do), it would find in this persistence the necessary affirmation of the impossible, or some literary *signification*.

But please note that, in the end, even when we compute with more than zeros and ones and accept both *alpha* and *nonalpha,* we normalize the intellective through its formalization. Then, would we get *the* cognitive schema of literature, everything *else* would yet be to expressed, created, and signified each time anew. Perhaps is it impossible to keep some sense of discipline without trying to reduce the irreducible. Perhaps is it sometimes useful to show both the heuristic gain of simplification and its costs.

54.

The experience of sophisticated fictions takes place in the intellective space. I am referring here to complicated creations, requiring a level of cognitive expertise, often called *art.* There is, distinctly, an art of being a spectator or an art of reading as well. The level certain cognitive skills reach is, of course, a condition of access to the intellective and of what arises from there. Hence, different competences have different results.

I am attending a play: Medea is about to kill her children, or the main character in the final scene seems to be floating in the air like a cosmonaut, or Arturo Ui growls and barks in the prelude. I am fully immersed in those worlds, much more than I feel my body in the chair, even though I am aware that I am sitting in a theater watching actors playing. This is an old scandal for the cognitive. In his *Confessions,* Augustine wonders why we enjoy being sad and witnessing the demise of tragic heroes when we do our best to avoid misfortune.[86] The answer is unclear if it has to be formulated without contradiction. Precisely, the singularity of the experience lies in a significant impossibility: for us, all of this is for real, all of this is a fiction. This cannot be, but this is what we feel and think. Bertolt Brecht's *Verfremdungseffekt* is a technique of dissociation that insists on the disconnect. But no such "effect" is capable of ruining immersion. Conversely, the realistic illusionism of which Brecht was critical cannot prevent the spectators from apprehending the show as a fiction, nevertheless.

It might sound easier to say that, while attending a play, watching a movie, reading a novel, or staring at a picture, we are *alternately* enclosed in another world, and aware of the reality surrounding us, so that we are sometimes here and sometimes there. Yes, it is far easier. And this is the kind of description I want to eschew. Even with alternation, given that ideas are being produced through neuronal avalanches and parallel distributive processes in the brain, there is a mental overlap: what I was believing does not immediately disappear and is not exactly replaced by another mind-set. If I see Medea taking a knife while a neighboring spectator coughs loudly, I can certainly feel to be "almost" instantly expelled from the scene at which I was looking. There are also countless moments of other overlaps, where I bifurcate and continue to both seize the modified reality of this particular fiction and process the realness of my world. The art of the spectator or the art of the reader of sophisticated fiction is precisely to master a mental ability for diffraction, for the immersion has already to accompany a critical (or aesthetic) judgment *at the same time*. This time may well not be *exactly* the same, even though, *distinctly*, it is.

The few experiments addressing the cerebral construction of the fictitious and the real are not extremely helpful. Some scientists seem to be in favor of a complex construct with a large role devoted to the hippocampus, linking the creation and reception of fiction with memory (amnesic patients with hippocampal lesion having trouble figuring themselves in new, still unrealized situations). The precuneus might be involved, in coding more specifically for the familiarity of events (remember that *Verfremdung* is also *defamiliarizing*[87]). Then, the mental perception of fiction would proceed from varying degrees of binding activity in cerebral areas. The real could be *more or less* real and possibly troubled by a weakening of spatial and temporal organization (as could happen with the invasion of recollection in the present of déjà vu). In such hypotheses, we would sometimes be at the moving frontier of fiction and real. In this zone of contact, and through the systemic *différance* of brain activity, we are inclined to aggregate the separate of the experiential in "the same"; and we have to decide quasi-immediately if we rearrange the strangeness of the indefinite into mutually exclusive categories—or if we take them together, in the other weirdness of contradiction.

Games and plays are found across animal species. In most animals, these behaviors largely disappear with adulthood (but parents certainly participate in the plays of their children). A biological function of rehearsal and apprenticeship is probable, and lots of plays among young mammals involve situations whose features resemble a future made of relations of dominance, predatory attacks, and instances of flight. The scientific explanation does not go much further than this, and it leaves relatively open the question of why some animals continue playing when they grow up (self-domestication is a rather circular response). The conception of artistic fiction as make-believe is close to the biological approach. It pays no attention to development and presupposes a durable enjoyment in the mere pretense. This, I argue, is not accurate for understanding more than the sketch I improvise at the office with a friend, or my exaggerated tale of yesterday's adventure. Another level is reached with sophisticated fictions, through contradictory performance and lasting significance. I do not only recall *The Tree of Life*, I am haunted by it; I do more than remember this poem, I am possessed by it.

55.

Sherlock Holmes may not be the best example for fiction, despite the thousands of pages that have been written in Anglo-American philosophy about his address and habits. Our problem is not to delimit different worlds in communication with each other. It is not to assert a real world in which our fictions would take no part.

It is to dwell on the heterological line implied in our intellective comprehension of a work of fiction. None of this will be rendered as

$$\neg\alpha \land a\Phi\alpha,$$

where a would be an agent, α an object, and Φ an operator (for an agent a over some object α) of mental presentation. Such a widespread description of our participation in fiction furiously looks like a reduced version of belief, under the condition of falsehood (or nonexistence).[88]

If we were to construct an acceptable operator for the expert intellection of sophisticated fiction, we should take into account the possible modification of the categories of *belief* and *knowing* (B and K in the "doxastic" and "epistemic" logics of analytic philosophy). For instance, I know that the children on stage are not killed, but I also know they have been executed by their mother—I do not simply *believe* they are, while I also hold this belief. Or I know the actor is not flying, and that there is a trick, but I do not "believe" he is up in the air; I both consider the illusion and refuse to reduce what I experience to it. (This is the power of prestidigitation as well.) If we introduce an operator Φ relating to my singular mental disposition (i.e., to the way I "phil" under fiction), it should not be a mimicry or caricature of B or K—unless we try to fictionalize the logistic of analytic philosophers, a task I might be taking on right now.

So if I had to come up with a formula, I would rather propose

$$a\Phi\alpha \land a\Phi\neg\alpha.$$

Traditional rationalists could still argue that only "a part of us" is thinking this way, and *another* part a different way. For them, a would refer to a potentially open summation of independent sub-agencies $a_i, a_j, \ldots a_n$ or to a relatively permanent self being actualized differentially (with ai "expressing" a at a moment ti). Then, they would correct the previous equation into

$$a_i\Phi\alpha \land a_j\Phi\neg\alpha.$$

One step further in the same direction would be the appeal to the so-called possible worlds. A "set of all the worlds" would comprise "the actual world" @ and its counterpart w given by a particular fiction (including by counterfactual conditions). What a neat solution, when different worlds "access" each other through a binary relation R . . . We would get two distinct and mutually noncontradictory statements: at @, a could have the value 0, while holding a value 1 at w. The operation of fiction would just help an agent "phil" that something which is not could well be. *Cum grano salis,* the initial formulation might even be rewritten this time as

$$a\Phi(\Diamond a) \wedge a\Phi(\neg\Box a).^{89}$$

Alas, this "brilliant" solution makes us lose the specificity of Φ. More regrettably, we simply converted fictions into *possibilities,* for the sake of consistence. This was already Spinoza's proposal, who wrote in the seventeenth century that a "fictional idea" corresponds to a "*possible* thing, whose existence does imply no contradiction in virtue of its own nature."[90] Enough with this.

We do not only have two distinctly separate cognitive operations that we retroactively contemplate with dismay: we *phil* the difficult contradiction in the neuronal integration of these overlapping thoughts. My *critical absorption* into a play does not have to command a general attitude vis-à-vis all possible objects (even though it could be that, just after leaving the theater, the usual barriers of the real have changed for me). Then, Φ would indicate the mental ability for an agent to hold together incompatible or contradictory statements at some point, and in some conditions,[91] without entailing the fictionalization of everything. The intellective would encompass the high point of fictive art, as we create and experience it, through novels, poems, movies, and life.

The diffraction on which we commented is sensible in the theoretical gesture of separation between planes, times, instances of agency, and so on; it appears both as the goal and means of a reversion to the external border of cognition and as the inevitable (though possibly delayed) aftermath of contradiction. Thus, it is a rejection of the dialogic as well as a recognition of its transient advent. Formalized notations are linear tools that unearth these ambivalent gestures. A consistent approach stems from the rejection, a dialetheic or paraconsistent interpretation tends to mute the singularity of contradiction. At any rate, all these paths are of interest to us if we seek to describe the virtual amplitude of the intellective space—but *what they will never accomplish also situates our task.*

56.

As in a fiction, the intellective makes sense of making none. All of this is nothing, a background noise, some foolish impossibility. But I live and feel it, and comprehend it, beyond the descriptive capacities of my own mind. This signification is real: it animates me.

The discursive hypothesis of an intellective space is a tentative response to the discontinuities, discrepancies, gaps, and breaks of knowledge and reflection. A *response* is no "reaction," and it should not be seen as a redemption of the defective, for the lacks we encounter are as unsurpassable as they could be. I need to say it one last time: the negative modes we roughly identify under the category of the unthinkable can only appear through the experience of thinking, an experience that indispensably needs cognition at its core. The intellective is not a priori, and it has little to do with mysticism: the incommensurable is a consequence of measurement and computation. The differential and shared performance of cognition, especially as it nears incompleteness and interruption, gives way to the impossible possibility of *dianoēsis,* which we might decide to conceal and deny because this makes our intellectual and scholarly life easier—or to interrogate and produce in a way that will irremediably be tied to the dynamic of cognition and to extraordinary regimes of thinking, including the literary.

II

ANIMAL MEDITATIONS

57.

Our thoughts install us in a real fiction.

I wrote "a real fiction." Usually, only one part of the question is considered.

That we all live in the fictions created by our brains (and/or the government, the spectacular, discourse, etc.) sounds like a postmodern motto, and, in Philip K. Dick, Jean Baudrillard, or *The Matrix,* it certainly is. The older formulation, deployed by Arthur Schopenhauer, would make us prisoners of *representation.* Even more classically, we are said to often be fooled by our senses and our "customs." According to Plato's cave allegory, a shadow play is what most of us contemplate, for lack of frequentation of the *noētos topos.*[1] There are large divergences between times, styles, and doctrines here. Yet, we find an invariant core argument: the world we take for real is a simulacrum; the majority ignores it, though a revelation or a proper technique could grant us access to another order of the real (or even to reality itself). All of this happens through our cogitation. Via a reform of our own understanding, being refined or illuminated, we grasp that the way we normally think reality is biased. We are piercing through Maia's veil. Today the cognitive sciences, and the study of the brain in particular, have officially become one of the most "legitimate" tools for this kind of reform.

That said, is "fiction," once it is recognized as such methodically, in any way *dispelled*? I still see a broken stick in the water. I am still in love with this woman and not consciously computing the optimal conditions for the diffusion of my gene pool. I am still reading these words as if they were speaking to me. Adventitious conceptions could be abandoned, and the physical explanation of a miracle may conduct us to disbelief. The fictions structuring thinking through intense repetition, self-domestication, and cognitive acquisition are not easily driven off. Insofar as they are adapted to the exercise of our living mind, they will remain for long, and will not be instantly superseded or "eliminated." Banal episodes of altered consciousness already show us that the real is constructed—we "know" it, but how could we instantly and durably suspend the experience of our experience?

Furthermore, could we even determine that a reformed approach would not be another fiction? A logical answer would be to test the soundness of the putative nonfiction, by piercing it and going to the level $n + 1$, in the same manner we left normal illusion by accessing a higher level of understanding. If we do this, then our reformed comprehension is also a fiction (it does not encompass the whole realness of the real). And if we can't go further, then we have no *real* test of validity. A pragmatic response would consist in checking some effects of "fiction" and "nonfiction" before comparing them. The engineer's approximation could ultimately suggest that there is a more accurate representation but no absolute absence of fictionality. Conversely, all the fictions purported by our brains or our societies are even minimally *real* as soon as they are materialized. My wildest dreams are more real than is a tsunami two years before it happens.

A *real fiction* is both a fiction of the real and the real of fiction. We just need one moment of separation from what we *think* is real to get an impression of the unreal; from now on, we will have a doubt that will not be easily reduced, so we are going to entertain more bifurcated thoughts. The first fiction—one that, seconds before, could not be seized as fiction, and maybe never was *one*—did not vanish. It has been fragmented, both abandoned and kept "at the same time." The real fiction I now experience as through a Φ-operator is in its turn a fiction, a real one, and so on. The cognitive sciences do not make me *more* real. And they do not prove that I live in a mere dream.

The driving forces of cognition, as they have been acquired through development and evolution, have side effects on the content of (meta)cogitation, and these consequences do stay, despite newly obtained knowledge. What we described as the enfolding of thinking onto the cognitive—a resilience that the sciences precisely exploit—is in a sense preceded by another return to the implied representational order of reflexive thinking. A journey through the sciences leaves us with explanations on one hand and with displaced but enduring conceptions on the other. If, with the sciences (and not exclusively *within* them), we think about our thoughts, we will undoubtedly distance ourselves from common ideas, *without* entering another era of the real.

58.

The real describes where we stand. We cannot stop here and subordinate the phenomenon to the noumenon. There seems to be a reality "behind" the real, or an *inconstructible,* that my words and cogitations could only indicate and systematically threaten to apprehend through mental construct. It is reality that is inhuman and meaningless—until we come in and make it real, in part. Realness is the quality of our real account of reality. Realness, then, could be enhanced by the recourse to scientific explanation, though it will not close the gap with "reality." At a subjective level, realness is felt through the contextual adequacy of what comes to us from reality and how we react to it. Because some logic is implied in most of what we process, we enjoy the conformity of a sequence of events with our own predisposition to derivation. We can also measure the correlation between an object and the verbal description we deliver from it. In our civilizations, the predictability of success for mechanical operations is another, central way of establishing the quality of our account. The mutual comparisons between approaches are much more practically oriented than they are theoretically founded, and they obviously rely on epistemological choices. "Truth" is one problem among others and not necessarily decisive in this regard.

59.

Scientific or speculative realism can claim to reach a higher level of realness than other discourses, in proportion to their methodic assumptions. Realness relates to how well the real is constructed, with certain goals "in mind." But the very reality of the real is incommensurable. Realism: *cosa mentale.*

The real is a *cognizable cut* of what "reality" indexes: it does not come from nowhere, but it no longer belongs to what the *"Sache selbst"* seems to recover. *The real only* relates *to "reality."* It is, along with all our thinking, what Vladimir Jankélévitch would have named an *organ-obstacle.*

60.

We should stay firm on the incommensurability of reality vis-à-vis the real. Correlationism and all the more philosophical realism implicitly write (informal) *equations* describing the ways we can (or cannot) access the thing itself. Would R stand for *reality,* r for what is *real,* n for the noumenal construction of any object, and u for some unintelligible; would we try to write equations with those letters—none of them would ever be *for real.*[2] While we believe we should separate reality from the noumenon, we simply have no *idea* of what it could be per se. Is it a transfinite accumulation of *realia* or a common abstraction of the differential? A summation, or a greatest common divisor? In what sense could we even speak of "the real," and how would we transform this meaning into mathematical symbols? Then, none of these definitions would ever *equal* reality, or rather what this word makes us imagine what is, independently from "us." We can bet that no function ξ will ever allow us to write $R = \xi(r)$.

61.

Mathematical notation offers no particular guarantee of reality. Formalization undoubtedly impresses people who are not in the know. More benevolently, it may serve as an interesting *analogy* within a theoretical discourse. Or, in the sciences, it could recapitulate, stabilize, and strengthen an argument. The use of an equation or a geometrical figure first confers a specific process of verification according to rules of automatic derivation, without too much interference of the semantic. As such, this does not say anything about the accuracy or virtues of this use. These aspects have to be *tested* through the artificial creation of controlled conditions. The *"adequation without correspondence"*[3] of computations and predictions with contrived and simplified portions or situations of the world, what does it hint at? It suggests that a mathematical description was justified—not that it transcends nature. Although the cognitive objects of mathematics are supposed to be *tenseless,* not only do they change throughout the history of mathematics (this is still a minor point) but their adequacy to *phusis* is *local, approximate,* and *temporary.* The adaptation of mathematical internal truths is local, as exemplified by the gaps between the level of physical descriptions and implied by epistemic incompleteness. Approximation is everywhere and begins with the parameters that are ruled out, for the sake of a demonstration. Scientific revisions, finally, will relativize or even void what once was assumed to be a perfect adequacy. A multitude of perfectly sound mathematical equations could be applied to a physical system, some of them being to no avail, and others representing an acceptable description, in the current limits of the known. Nature is not written in mathematical language, but we can write equations or form figures about the construction of *phusis.* We are always *testing* mathematics: through writing and drawing (as Brouwer suggested[4]) and through a confrontation with the real. *Reality qua reality* is "realized" through cognition, observation, selection, delimitation, experimentation—and occulted "as such."

The fascination with the heuristic potential of numbers began early in the theory of nature. The ancient Egyptians had a mystique of arithmetic that was transmitted to Pythagoras, then to Plato, and so on. The *Tao te ching* also explains that the "unnamable" "*tao* engenders the one, the one engenders the two, the two engenders the three, the three engenders the million of things."[5] We cannot conceal the depth of numeration in our cognition: it is biologically more ancient than language (and more widespread among animal species); it also governs a lot of what we think. But this intellectual *organon* should not be the pretext for the idealization of the holy numbers. Moreover, *mathematics* is not given at all. Gathering geometrical abilities in space-time with a formal aptitude consolidated through verbal language, with numeration and logicality, then taking the whole as inconstructible "mathematics" ratifies a contingent disciplinary history, while feigning to find an absolute. As for the unification of mathematical science, the dream of formalists or even Bourbaki, it is as probable as the physical theory of everything.

At the antipodes of formalism, Brouwer's intuitionism, which "separates mathematics from mathematical language"[6] and posits the former as a "languageless activity of the mind,"[7] is compatible with some posterior insights about the animal cognition of numbers and space. And because, for "*pure mathematics,*" there is "*no sure language,*"[8] paradoxes or undecidable antinomies are perceived as mere effects of the language-like structure of science, endowed with symbolism, semantics, and syntax. Brouwer's maximalism (all mathematics should conform to intuition) engenders some practical minimalism (entire fields of inquiry or the principle of the excluded third, for instance). This was an evident obstacle to the acceptance of his ideas. Furthermore, even if intuition comes first, why constantly come back to it? Then, how could we be certain of the maintained purity of mathematics outside of the mythical nonverbal and contemplative subject or through its indexical transmission via symbols? Yet Brouwer may have been right to find within a property of the mind the *impetus* for what became a discipline. What he missed—in his analysis of the primordial *intuition* deriving from "the perception of a *move in time,* which is the falling apart of a life moment into two distinct things"[9]—was the fundamental link, acquired though evolution and adaption, between our noetic structures and the scale-dependent regularities of the universe. The diffraction of the *now* (giving way to what Brouwer named "the two-ity"[10]) as well as the impossibility of such a bifurcation (reinterpreted as "empty two-ity"[11]) configure the real for us. Being abstracted from our animal belonging to this world, this construct touches on "reality," though it can neither help *re*construct it nor stay in it.

62.

The final mathematization of reality is science fiction. The mathematical language of "Being qua Being" is another fable of the real.

So, what could we do with statements such as "mathematics = ontology"? What Badiou's pseudo-equation means in his work is rather unstable, despite what he and his official commentators could say. In *Being and Event,* the author warns his readers that "the thesis that I support does not in any way declare that being is mathematical. . . . It is not a thesis about the world but about discourse. It affirms that mathematics, throughout the entirety of its historical becoming, pronounces what is expressible [*dicible* in the original] of being qua being."[12] Staying true to his pseudo-equation, Badiou mostly delivers, under the name of "metaontol-

ogy," a philosophy of mathematics. In this perspective, he introduces, comments, and reflects on formalized propositions. He
does what seems to me a quite partial reading of the power of
contradictions—such as $\neg(a \in a)$, which he calls a "paradox" or "a
joke"[13]—and from there, he derives a set of theses about the "void"
of being. But all of this is arguable.

Now, Badiou also opts for a second regime of relation to the
mathematical. In the same way he uses verbal language to add
what he sees as a "meta-level" to an utterance of set theory, he additionally transcribes into "equations" what he believes to be the
core of his discursive argument. A subjective and present fidelity
to an event will, for instance, be written as the following "mathem":
"$\varepsilon/\not\subset \Rightarrow \pi$"[14] The justification of this mirrored gesture of inversion
(my words can comment philosophically on a mathematical nota-
tion; an equation can "recapitulate" a verbal theory of the subject)
is simply unwarranted. The two thick volumes composing the dip-
tych of *Being and Event* are unable to intelligibly authorize this gesture and to make sense of this "mathematics of appearing."[15] There
is probably in Badiou a quest for legitimization and a reiteration of
Lacan's own recourse to topology, set theory, or graphs. Well, you
never know with Lacan. His work functions as a kind of chronic
hoax (or extended *canular*, to use a French word). The psycho-
analyst spoke of his "ideal" of "integral transmissibility,"[16] but he
also expressed a distrust in the autonomy of his mathematical language, as if those formulae were not sustainable by themselves (and
therefore mere illustrations).[17] In any case, each time conceptual
discourse is supposed to translate itself into mathematics, we may
stay within the analogical.

The other main problem (amplified in Badiou's 2006 essay) lies
in the subtle—but constantly repeated—shift from a discourse on
a discourse on being (philosophical metaontology) to direct assertions about the "level of pure being."[18] The caution and focus on the
"dicible" are by and by relegated to the background, giving more
weight to a philosophy that, as Deleuze once said, is in fact "the
return to an old conception"[19] and to the posture of truth teller:
this appears to be the main "reality" of "the" concept according
to Badiou. Deleuze was inaccurate on another point, however:
Badiou's work is not "complex,"[20] it is *complicated*. Mimicking the

dogma of the one being and of the appearing multitude, Badiou begins with very simple premises then accumulates overabundant details. This makes the academic assimilation of his thought quite straightforward at its core (a few doctrinal items need to be learned by heart, such as the pseudo-equation), while allowing the scholarly pleasure of transfinite pico-discussions about lemmas and commas.

63.

On the intellective space, real fictions are apt to be contradictorily seized, instead of being rejected, linearized, or gently combined into dialectics (be it negative), as ultimately happens in the cognitive. The real is both less and more than itself, and our fictions of it also inform what and how we can think. By holding together the real and the nonreal, we approximate the unnamed coherence of the nonconsistent, or what the say of the incommensurable could express, like the illegibly significant imprint of reality.

64.

The more we observe animal life and the physical world, the more we are convinced that the human exception is a fantasy, quite explainable—and meaningless. In Schopenhauer, the discordance of *will* and *representation* is a tragic affair, ending up in neutralizing quietism. In Bergson, an élan vital has to be added to the scientific, because the latter fails to account for every dimension of our *noēsis*. Consequently, there is "necessarily some cerebral substratum for the psychological state" but "nothing more."[21] Pessimistic dismay and spiritualism are two *philosophical* attitudes, where the refusal of a Φ-operator logically derives from the alleged superiority of the principle of noncontradiction over *cogitandum*. On these points, the theoretical primacy of the cognitive is patent—even with Bergson.

It is true that our mental structures and the extendedness of intellection will, sooner or later, make us lose sight of the beyond where we bypass ourselves. But it falls on us to appeal insistently for an experience of thought that is neither a *pure* fantasy nor a metaphysical flux. In the same manner that we continue to designate defectively reality without being able to break away from the fragmentation of the real, our task is to produce the conditions for an expansion of the transient intellective.

65.

If "reality" is absolutely absolute (i.e., freely independent and *separate*), no description of it (including a mathematical one) is more than real. Quentin Meillassoux speaks of giving "the intellectual intuition of the absolute,"[22] and this is precisely where he stops: we can only *intuit* the nonreal of reality. Through the experience of the disconnects of the real, we feel the necessity of a "Great Outdoors." The moment we think about it, we are in the real and lose the unthinkable absoluteness of the absolute—that we never touched anyway.

There is also an *intellective* intuition of reality, and of what it could do to us, through the separation of thought from itself.

Descartes asks, Where could my idea of God come from, being myself human?[23] The idealization of our ideas simplifies the relative points of stabilization in the dynamic of thinking, and it reflects a perceived stability of the macroscopic. As for the category of the absolute, in all its guises, it now appears to us as the retro-interpretation, by the cognitive, of noetic fragmentation and of the possibility of intellective journeys. When the defective is not marginalized or omitted by the enfolding on cognition, it is kept as a distorted sign of Reality. There it is conceptually radicalized (the separate becomes the absolute), repaired (the cognitive refection of the defect is taken for perfect), eternalized (the transient is the source of all time and duration), and unified (the singularity of the *tracé* is made the One). The idea of *God* is a cognitive consequence and appropriation of the intellective. This makes me doubt that horses could ever have deities, as Xenophanes ironically proposed.[24] Though, certainly, all the animals sharing the extension of intellection could imagine godly figures. The gods, in their narrative existence, are here to make sense of the no-sense; they extrapolate some aspects of the real. As theoretical concepts—be they named IHVH, *to theon,* mathematics, or Nature—they requalify the narrated in combination with a deceptive acceptance of the supplemental space of *dianoēsis*; they mytho-logize reality. The divine becomes the main attribute of reality, once the cognitive discursively thinks it *is* the intellective. Spinoza's *Deus sive natura* is an expected point of arrival.

Conceptual gods resist rationality, for they come back from the trip beyond—and they perfect it, in consolidating the cognitive rendition of what escapes it. They are not opposed to human reason, which made them its corrected image. They are ambiguous objects, impossible to treat philosophically or scientifically—and still doomed to haunt the most positive discourses, over and over again.

66.

The Vienna Circle put forward that any consideration about the reality (or lack thereof) of the external world was made of "pseudo-statements." In this respect, metaphysics as a whole was nothing more than a long and painful *flatus vocis*. I am almost tempted to agree. On one hand, there is certainly no way the human description of reality itself, delivered in any form of language, would be a definitive and positive statement. And on the other, metaphysics is a *discourse* through which one tries to think; it is neither the *deixis* of the empirical nor an analysis of possible assertions, which is why, even in new clothes, it merely contains and conducts intuition, without furnishing the demonstration it claims to deliver. But, as far as I am concerned here, two major errors are committed by the Viennese promoters of an "elimination of metaphysics": one is about the alleged "meaninglessness" of pseudo-statements, the other is the depiction of the only useful philosophy.

Carnap is exemplary. His conceptual understanding of meaning is as restricted as possible. In his 1932 article against metaphysics, Carnap admits that a word has scientific meaning "by reduction to other words,"[25] apparently seeing no difficulty in the limits of definability and the movement of internal regression. The example of "arthropod" is pretty comical, for the corresponding basic "observation sentence" "*x* is an animal," though practically pretty straightforward, is scientifically unclear. Saying that the "relations of deducibility to the protocol sentences are fixed, *whatever* the characteristics [and, I assume, the definability] of the protocol sentences may be"[26] is preposterous. When the stratified construction of meaning fails, a constant reliance on the empirical is presented as a solution—the empirical playing the structural role of "reality." Apart from the "empirical criteria,"[27] the other feature of meaning—deducibility—is cast in bivalent standard logic. However, for Carnap, nonscientific empirical meaning could be found in mythology or poetry. He estimates that the word "god" in Homer "has a clear meaning"[28] because, in this case, it "is used to denote physical beings which are enthroned on Mount Olympus, in Heaven or in Hades."[29] More generally, all metaphysics "is based on a logical defect of language,"[30] a defect that is corrected in formalized science and at least attenuated by the empirical. The problem is that the word "god" is semantically diffracted in the Greek *epos* as well, and that our logical lines might be broken too. Carnap puts metaphysics on trial to keep a "logical composition of the world" and preserve a relative consistency of language. By restricting meaning to the logical derivation between words in an empirical context, Carnap fails to acknowledge that the trouble with metaphysics is the symptom of a larger and more profound defect.

The "new" philosophy of the Vienna Circle (a *method* of logical analysis escorting science) avoids the transference of the sacred and the semi-hidden conservation of the divine. Unfortunately, this welcome movement of secularization and emphasis on both meaning and experience also kills thinking and pleads for the scholarly rule of the *cognitive*.[31] However, some nostalgia for the poetic remains in Carnap. It is obviously anchored in a fallacious view of the literary as a mainly expressive (and nontheoretical) art. The nostalgia is still expressed and even includes a rather reverent allusion to Nietzsche, a point that has been quite systematically overlooked by the authorized commentators of the analytic tradition. Did Carnap see, despite forceful narrow-mindedness, the possibility of the intellective? By ascribing a beyond to poetry (even if ill-conceived), he may have said something differing from itself, something cognitively inaudible, especially to the analytic partisans he was lining up. Now, the project of poeticizing the metaphysical could only be performed through a detour via the intellective space, which is against both "metaphysics" and any decorative or "purely" emotional version of literature. The poetics I have "in mind" is no Heideggerian oracular originarity, no prerational outburst of meaningless affects; quite the contrary, it speaks afterward, then takes into account the trajectory of defection.

67.

Aristotle's attempt to build a "first philosophy," later called *metaphysics* by the tradition, mainly resides in its definition. The first philosophy is said by Aristotle to be "theological,"[32] which opened the way to medieval scholasticism and, from there, to the historical companionship between metaphysics and discourses on gods. Aristotle's question was rather the *divine*,[33] which does not need to be an attribute of Zeus, Jesus, or Allah. The divine recovers the "first principles and causes"[34] or what makes nature *(phusis)* what it is.

But we have two distinct problems. One is, What makes reality what it is, that is, reality? And the other is, What makes reality real? The second interrogation has one answer: our mind. The first one has none—at least, none that we can comprehend or express. An explanatory response (physical laws make *phusis*) is ineffective, and as soon as it *means*, it displaces the first question toward the second. If reality is taken for real, all the considerations we made about thinking were necessary—and they suggest that the real is both stable and mobile, common and shared, and that it makes some sense to us, despite the disturbing intuition of the no-sense of reality. Aristotle, with a large part of the metaphysical tradition he allowed, conflates the two questions. He presupposes that, because we are real, the shape of our thoughts is determined by principles that also affect reality as the unrelated nature of what is. One usually forgets to mention that his quest is for the principle of "being qua being" (if we want to stick to this common translation) and of what allows being to be "by itself" and "according to itself."[35] This is how the principle of noncontradiction is both *de re* and *de dicto*,[36] to use scholastic distinctions that analytic philosophers still enjoy. Of course, a contra-*diction* is potentially *de re,* if we accept that the real is constructed; but it cannot concern what "reality" indexes, so this leads to aporia. We also admitted that our logical apparatus, and particularly those parts that seem to be in common with other animals, was plausibly based on observable regularities of the macroscopic world. Here logicality is not unrelated to reality, even though it obliquely proceeds from it and cannot help retrieve it per se.

"What makes reality real?" is mainly a question about how we think, through, with, and beyond cognition. We have multiple and repeated intuitions of a discontinuity between reality and the real and no ability to think the former without constructing the latter *instead*. We can argue that our thoughts are configured in such a way, both by evolution and contingence, that the "natural structures" of the mind give us some image of reality. And then? We cannot go from an equation, a piece of apodictic prose, or an insight *to* reality. The incompleteness of our thought may be that its reality will just be real to us. A scientific understanding provides us with real fictions of reality. Going further means attributing signification to such fictions on the intellective space—and neither unifying science, which can only be expanded at the same pace as nescience, nor synthesizing reality, which is inconstructible.

Metaphysics would no longer be a causal discourse on the first principle of reality—but, in the ultimate defects of the constructed real, it would dialogically locate the *tracé* of reality.

68.

Reality *allows* the real through diffraction in the finite processes of cognitive life. The commonness of the *organon* (from the eye to verbal language) and of the effects of reality gives us a series of real pieces that we can exchange and share.

69.

The trial of metaphysics in the twentieth century developed three major arguments: the modern study of nature (physics) has invalidated the need for a "first philosophy"; metaphysicians do not even know what they say or mean; and their discourse has been recurrently shown to be heteronomous (to religion, to politics, to subjective desire, etc.), which ruins its pretentions to absoluteness. This critique was articulated by philosophers who were completely adverse to each other and were living either on the "Continent" or outside. It has been extremely successful in officially discrediting a subdiscipline. Nonetheless, it did not *eliminate* metaphysical reasoning (disguised as social science, for instance), and it did not even prevent a return to the label of metaphysics, as seen in the scholarship of the last decade in particular.

I doubt we need a "new" metaphysics, coming after its own indictment. But if there is any hope in this direction, such a discourse should be inscribed in the lack of physics. This implies that only a reflexive study of "nature" could make sense of what is not reached or understood. Then, neometaphysicians would not know more what they mean than their predecessors did, but they might try some ways to make their texts more *significant.* As for the absolute, they could not find in it more than a mark of separation from reality. Instead of assigning to the metaphysical a place above nature, its locus would be beyond the practice of all "physics," or beyond the cognitive matrix, or where we intellectively intuit the disjuncture of reality and the real.

What I suggest is rather a "last philosophy," in the defection of the conceptual. I may be closer to what Alfred Jarry once called *pataphysics,* or "the science of imaginary solutions,"[37] all concerned with epiphenomena that are both more and less.

70.

One more time. We can imagine that *nothing would be real without reality* and that *reality is missing from the real.* A metaphysics of the real could be developed as extended physics—encompassing, within the study of nature, the explanation of constructive cogitation—from which signification would arise. The construction of the real (which leads to its deconstruction) limits the metaphysical so drastically that the old proponents of first philosophy may refute the recourse to such a vocable. This neodiscourse, however, would come *after (meta)* physics and stand in its significant *beyond (meta).* This is what I tried to give in these pages.

That reality is missing from the real it also allows suspends our faith in a discursive or noetic leap toward reality. There is no absoluteness of capital Metaphysics. In this direction, we may find at best a handful of metaphysical intuitions. In particular, both the *scar of reality* in our thoughts and the *phúsis* of our survival could make us believe that the constitutive lack is not without concordance with the defectiveness we observed throughout our inquiry. This would be said as "everything that is fails to be" and should be understood as a judgment on the real, trying to reach further. If I had a major metaphysics of reality to constitute, this is where I would begin. But no crystallization would be possible without ruining the hypertranscendental claims of my reflection, so I would be left with my imaginary science.

71.

This happens over time. The physics of being describes dissolution, decay, diffraction—at some point. "Being qua being" will no longer be "according to itself." In speaking of "nonbeing" for what could be, what will happen, and what no longer is, we are constructing a logical category whose mission is to inscribe the world in bivalence. That being would cease is much more perilous for any metaphysics than the considerations on $a \wedge \neg a$. We are facing another difficulty: something real *fails* to be what it is, in virtue of its construction. It is less than it is, or than its "is."

In the wake of its own saying, the intuition of failing reality could not be completely maintained as well, leaving the impression that the defect could be defeated. But how? By the fact that something tends to be, nevertheless? We'd border the indeconstructible tip of deconstruction in Derrida, the "?-being" of Deleuze, or even Meillassoux's "maybe." But, in these last two manipulations of "being," I fear we have already come back to the real, instead of keeping with what the word *reality* awakes in us. Is the failure of being, while something is, another enactment of disjuncture?

72.

Inconstructible reality is felt each time the defection of the real is incomplete in its turn. Then, in its singularity, a metaphysics of the real does more than explain the construction of nature or thinking cognition. It enriches our theoretical imagination. As a discourse of the metastable, metaphysics is a vision of the mind.

73.

"The mind is limitless and self-ruled, and it is mixed with no material thing . . . , for it is the thinnest of all material things, and the purest too, and it holds all knowledge about all things."

In those lines excerpted from his treatise *On Nature*,[38] Anaxagoras elaborates a physics of the mind. This material mind is the principle of what is, and it immediately installs a structure of intelligibility that we, humans, depend on. *Nous* (mind) here is supposed to be at the core of reality. Anaxagoras's influence is crucial on both Plato and Aristotle, who retain the motif of an "intelligent design." The motif of a "soul of the world" is largely developed in the *Timaeus*. We recognize in Anaxagoras's cosmic mind a retrospective interpretation of human intellection, suddenly promoted to a godly status, through dehumanization (*nous* is no longer bound to the animals we are), conceptual repair (it is "limitless," "self-ruled," and omniscient), and hyperbolic celebration ("the purest," "the thinnest"). The intelligibility of reality is assured by the presence of a tenuous and invisible mind pervading nature. Both Plato and Aristotle will abandon this materialism and posit that Ideas or "the god" are principally involved in the constitution of things, making the latter mentally penetrable.

I believe that Anaxagoras's suggestions could be adapted without any deep difficulty to the present context of mainstream cognitive science. One just needs to add "the brain" here and there or to assimilate "the thinnest of all material things" to electric and magnetic fluxes. In a comment he gave on Emily Dickinson's poem "The Brain—Is Wider Than the Sky," Steven Pinker is very seriously (and unintentionally, I'm sure) repeating the position of the pre-Socratic philosopher. Pinker notices "the grandeur in the view of the mind as consisting in the activity of the brain."[39] He admits that "science is, in a sense, 'reducing' us" to a hunk of matter" but maintains that "the brain" (and please note the recurring use of this term in an absolute way, a tic we find all across the board, from Pinker to Malabou) has a "limitless ability to imagine real and hypothetical worlds."[40] That being said, Pinker concedes that "the mind, in contemplating its place in the cosmos, at some point reaches its own limitations and runs into puzzles that seem to belong in a separate, divine realm."[41] In conclusion, the brain–mind is "perhaps as weighty as God."[42] This is where "the scientist" (?) stops, while remaining convinced of the *natural* intelligibility that allows his own visions.

In this hodgepodge of (meta)physical considerations, the limitless mind collapses on "its own limitations," the divine is both source and effect, the animal brain is subsumed under the Holy Brain, and sky and water are *as such* within the brain ("the brain of the reader must contain the sky and absorb the sea"[43]), reflecting that nature is organized by (a/the?) transcendental Mind. The inner discrepancies, as well as some blatant contradictions, are the obvious mark of a hectic journey on the intellective, which a dogmatic reversion to the cognitive clumsily attempts to conceal. We lose the philosophical strategy of Anaxagoras and his followers because Pinker's reflection is situated in the explanatory plane of science and marred by his lacking understanding of signification— and human thought. While the goal was to reinterpret the real (including neuronal processes) as scientifically given, the result is a marriage of metaphysical biases, partial repetitions of conceptual discourses, and delusions of grandeur.[44]

The cognitive encomium of the Sacred Brain is a perfect ally to the widespread description of the "intelligent universe," a cosmos of physical information where thermostats are endowed with experience[45] and the ecstatic—and ill-named—"Singularity" of the transcendental human will know all.[46]

74.

> The Brain—is wider than the Sky—
> For—put them side by side—
> The one the other will contain
> With ease—and You—beside—

The Brain is deeper than the sea—
For—hold them—Blue to Blue—
The one the other will absorb—
As Sponges—Buckets—do

The Brain is just the weight of God—
For—Heft them—Pound for Pound—
And they will differ—if they do—
As Syllable from Sound—[47]

In her poem, Dickinson uses a series of *adunata* (putting "side by side" the Brain and the Sky, holding the former with the sea, hefting it with God) that should be performed nonetheless. The *I* that I am (the "you" appearing in the first stanza) has the ability to respond mentally to the poet's injunctions and thereby *measure* and *test* the validity of the statements delivered by the first line of each stanza. I can only do this if I accept that "the Brain" (and not "my brain," or even "the human brain") is as unique and circumscribed as "the Sky," "the sea," and "God" are. The physical measurement that the text invites us to perform repeats the *objective* part of science through a description of width, depth, and weight. This is the human formation of the real, which is why *I* (cf. "and you") am also *contained* within (l. 4). This measure is also profoundly defective, for it is unable to produce its object of reference: not only "the Brain" is made real by its severance from the organ (or it would not be "wider than the Sky") and has to be converted into its activity, but it is hypostatized through its concept. Then, the disjuncture of reality and the real, in the production of thought, is inscribed in the poem. Although Dickinson's heavy recourse to dashes is not specific to this text, here it suddenly and repeatedly interrupts the assertive tone of the text. Aphoristic authority (expressed in the first pseudo-metaphysical statement of each stanza) and logical experimental evaluation (introduced by "For," which follows the imperative mode of a scientific protocol) are verbally evoked and concurrently revoked by the hiccup of the em-dashes. The grammatical construction "The one the other," appearing twice, hints at an identity (through absorption or containment) that the comparative ("wider," "deeper") seems to prevent. The line "And they will differ—if they do—" accomplishes the dialogic of the poem, bringing up a reflection on the intellective space of language: differences between "Syllable" and "Sound" are nonexistent outside of verbal semanticism. Inside literary discourse, they are requalified by signification. Moreover, the uncertain differences between God and the Brain, between reality and the real, between a cosmic *nous* and a human *mind,* are nondefinitively settled through language.

75.

In pretending that "of all the material things, the measure is man,"[48] Protagoras was apparently taking Anaxagoras's argument one step further, while suppressing its motivation and repatriating *nous* to the human mind. If we believe the tradition, Protagoras's motto was used in a kind of interested pragmatism, where truth was built on the needs of the moment. This is not what we are looking for. A few revisions may yield other meanings. The main problem is with the word *metron.* Man is not the *measure* of all things but the *measurer*: the one that does so with his (extended) mind. The measuring standards are constructed by humans, under the pressure of what is inconstructible to them. The inconstructible is both measurable and immeasurable, being in one case potentially reconstructed (or unmeasured as such) and, in the other, made incommensurable to us.

"That which does not depend on us" also includes *us* to such an extent that we cannot abandon for long the standing of the real in the development of our discourse: mathematical description or object-oriented imagination will not be the proper solution. Yet, if we want to think beyond cognition and experience the self-ruling automaticity of our defective mind without limiting ourselves to it, we have to reconsider our animal *ethos.*

76.

Throughout its history, the science of mind has been dominated by a vision of its object as something divine. Most divergences of approach were about the limits of limitlessness, the actual range of self-rule, and materiality. The more secular version took the divine as the perfection of computability. Quasi-limitlessness became the mystical number of a hundred billions of neurons; self-rule was translated as automaticity; materiality was emphasized. All in all, the sacred was slightly downsized, but the sheer perfection of automatic, logical, and mathematical processes sounded a lot like theology. Even when defects were acknowledged, the defective was neglected.

Moving toward intellective studies would require that, *as much as we can,* we accept ourselves as constructing and constructed animals. We are conducts for chemical and physical reactions, material packs of cells, living bodies, communicating creatures, operators and prostheses of transferrable cognition, organisms embedded in semanticism, performers of intellection, interpreters and creators of differential ideas. Self-organization is not excluded (is it a demiurgic rest?), the inconstructible is not denied, but they make sense with our cogitative animality.

77.

Modern science has powerfully advanced that humans were animals. This should be considered as a decisive conception and a piece of the *flourishing* metaphysics Charles Darwin was promising to himself through the understanding of baboons.[49] Conversely, this scientific "discovery" has always been a kind of given in most civilizations. The obfuscation of "our animal nature" is just a local phenomenon, encouraged by some monotheistic religions and their secularization through the technology of the divine machine. The "news" is old. Even the Greek thinkers were using *zōon* when referring to *anthrōpos* (human) in a larger frame: a *zōon* is *something living* (by etymology) and an *animal* (by usage). There was no shortage of qualifications for the human animal in classical Athens. Two general ideas were so widespread that they were rarely formulated: humans are animals, humans are immediately recognizable by humans. Hence saying that *anthrōpos* is like this or that does not imply that this is its only, or even most crucial, feature. The implicit, of course, is also apt to entertain xenophobia and makes it easier to situate slaves or "pygmies" on the margins or the outside of humanity. One finds several qualifications of the human in Aristotle, each one having served as autonomous *definitions*. The anecdote of Diogenes the Cynic, coming to Plato's school with a plucked chicken and saying "here is Plato's man (human)" to ridicule the philosophical definition that was taught at the Academy,[50] reminds us of the perceived impossibility of an accurate description as early as in Greek antiquity—and of the conceptual necessity to explore nonetheless what humanity could be.

78.

Decades before both Aristotle and Diogenes, Sophocles makes a list of human qualities in the second chorus of his tragedy *Antigone* (possibly performed around 441 B.C.). The first line of the chorus is justly famous. It says, in the old translation of the Loeb Classical Library, "Many wonders there be, but naught more wondrous than man."[51] As has often been remarked, the terms given here as "wonders" and "more wondrous" (in the original, *ta deina* and *denoiteron*) also conjure up the ideas of *terror, frightfulness*—and of *skills* and *cleverness*. The rest of the chorus should be heard according to this significant oscillation. A movement of characterization of the human opens up a paradoxical praise, where the cleverest inventions could be used as means of destruction: the last antistrophe makes clear that "skill" or "wisdom" (*sophon*[52]), if individually deployed against the gods and the City, will not serve well. Because we are moments before the apparition of Antigone, who defied the written laws, and before her controversy with the king Creon, who promulgated the rule, one could say that the ambivalence of *deinos* is about to be simplified by the plot itself. But tragic choruses are not only steps into a diegetic progression; they also function as instances of general reflection and are regularly entwined in a dialogue with religious discourse, sophistic or philosophic argumentations. The *ingenious* human is, at the same time, *terrible* and *marvelous*. All the contemporary rhetoric of technophilia and technophobia with regard to mankind is inscribed in these lines—but it appears in Sophocles as an intellective contradiction and not as a rational di-lemma.

What are the achievements of the human, according to Sopho-cles? Agriculture appears first, followed by hunting and fishing, as well as the domestication of wild beasts (here the bull). In the second strophe, "voicing wind-swift thought" is mentioned in con-junction with social and political organization, on which the last antistrophe will dwell. What accounts for all these feats is the intel-lectual capacity of men (their cognitive success, could we say?). The human is able to "teach to itself" (*edidaxato*[53]), which permits both knowledge of nature (especially the elements and the weather) and wisdom, however short it may fall when facing death. The acquisi-tion of knowledge and skills derives from what may be the principle of the human "ingenious (and terrible) wonder" (i.e., *deinon*): his techniques, arts, and machines, alias his *tekhnai*. The word *tekhnē* is used once, whereas *mēkhanē* appears three times, directly or in composition. *Mēkhanē* is a disposition of the mind, its ruse or *me-chanics,* and a tool or a *machine.* A wide array of agricultural tools, by the way, is duly mentioned in the first strophe and antistrophe.

Then, the ambiguous and unique wonder of mankind comes from the following features: a mental and practical aptitude to make tools (used in agriculture, fishing, hunting); a capacity to domesticate wild species; the communication of thoughts through speech; the ensuing organization of society; the acquisition of new knowledge about the real; and the ability for reciprocal teaching. These traits *wonderfully* correspond to the "new" approach that modern science brings us. In a recent text, based on advances in experimental psychology, biology, and paleoanthropology, Pat Shipman synthesizes the definitional question of the human an-imals in these terms: "Traits often considered diagnostics of hu-mans and significant in their evolution are (1) making and using tools . . . (2) symbolic behavior, including language . . . (3) the do-mestication of other species."[54] To this list, Shipman adds what she calls "the animal connection." Although there is some unex-pected vagueness with this fourth path, this suggestion could, in my opinion, be seen as a recognition of the animality of humans. In Sophocles's chorus, the word *animal (zōon)* does not appear. But it is the answer to the wondrous enigma about what "the human" is. *Anthrōpos* is neither a "wild beast" *(thēr)* nor an immortal god, the chorus says; but because he found no *machine,* and no *cure* for the fact of his "passing," he is simply *living* like a frightful *animal* (*zōon*). The whole chorus functions like an enigma inverting the one Antigone's father, Oedipus, solved as he entered Thebes. The answer to the riddle of the Sphinx was precisely *man* or *human* (*anthrōpos*). Here we begin with the solution (*human*) and grow back to the problem of the *animal,* whose name is both uttered and concealed. From the intellective space, we have to reconsider the puzzle and examine nondefinitive answers to the kind of animals humans could be.

79.

What modern science does, in regard to the human animal, is mostly to turn qualifications into definitions—none of them working as they should. *Are humans the only really bipedal animals or primates?* Yes, but no. They are not exactly the only ones, even among primates. The fabulous tale of Lucy, the first ape to stand upright, is gently outdated, and it is becoming quite obvious that, in certain conditions, orangutans or chimpanzees can walk on their two feet. *Aren't we Homo faber?* Yes, to some extent, our machinery is quite impressive and unmatched, but the use of tools by chimps in the wild has been documented from the 1960s on, and several other (nonprimate) species, including some corvids, are now known by human science to exhibit this competence. Kanzi and his half-sister Panbanisha, two bonobos reared in a Pan–Homo environment, have been taught by paleontologists to produce stone tools that are roughly comparable to the level reached in Oldowan technical repertoire and that they spontaneously use in a way that is similar to what the fossil records suggest. *But we have the largest brain, haven't we?* Almost (we need to exclude elephants if we aim to win this game). As for the cognitive advantages the size, configuration, and biological specificities our brain allows, they are difficult to discard. The problem is that, as we said before, fluent speech—which could logically appear to be the most immediate consequence of the history of our peculiar encephalization—came in millennia *after* the biological constitution of *Homo sapiens.* This discrepancy reflects the gap between our undefined understanding of the human and the anatomical and natural object that is our species. *So humans are animals endowed with language.* This is the strongest characterization. Now, we see exceptions to the rules: human infants, who have to be taught how to speak in the limits of the "critical period"; adults with language impairment or deficiency; other animals who participate in verbal speech through understanding and word mapping or even, in the case of languaged apes, by producing short and semantic sentences. Similarly, the symbolic is partially transferrable to nonhumans, and some apes have been very proficient at using masks or adornment. *What about politics, then?* Well, that politics is highly developed among contemporary humans cannot hide that a sophisticated ordering of social order is in place in many other animals groups. And even *domestication* is coming to be seen as a biological process that does not have to be human driven and could derive from an inner-group dynamics, under given ecological circumstances: some wolves would have been self-domesticated, hence more familiar with humans, before being adopted (and consequently changed) by their lives with *Homo.*

This list could go on and on. There is not one absolute characteristic. And no definition is adequate. In his last works, Derrida already underlined some major consequences to this situation. The intense circulation of motifs between poetry, philosophy, politics, and the sciences could be seen as the persistence of metaphysical assumptions up to the realm of the positive, or of a yet imperfect naturalization of the object *Homo.* Both interpretations may be correct. There are at least three other options that I would like to favor. First, the incompleteness of such definitions, *in loco humanis,* may be another limitation of our internal knowledge; this would call for multiplying qualifications and abandoning the idea of a clear-cut intersection between them, because they may operate on nonsecant planes. Second, the stake, that is, the *phusis* of human, cannot be *simply* decided by (meta)physics or any other discursive mixture. This shows that, for us as epistemic agents, the construction of the real needs to be interrogated if we want to make any sense. The anatomical and descriptive category of the "human" cannot be enough today, not only because it is already discontinuous (it is) but also because it presupposes nonanatomical (but biological) features in the circumscription of its object. Now, there is no reason to make an exception for humans, which should reversely imply that the scientific "naturalization" of the living object has to involve its conceptual construction, as it is conducted by ourselves, and, if applicable, by the members of the studied species as well. The biotope of humans includes their minds and so is potentially the case of many other animals. Lastly, the disciplinary and discursive core (from literature to experimental psychology) is no longer to be construed as mere evidence of a "lower" degree of scientificity but as a mark for a theoretical need toward the expansion of the intellective.

80.

When Aristotle wrote in the *Politics* that the human "animal" was the only one to have "logos,"[55] he was not the first Greek to link "humanness" to the faculty of language. Sophocles was already arguing in the same sense and summarizing a tradition that predated him. Isocrates, a former student of Socrates, also developed a similar thesis.[56] One could find equivalent claims in non-Western parts of the world. It remains the case that the way Aristotle framed the debate was extremely powerful and is still perceptible today. "Logos," in his text, bypasses communication. It refers to discursive rationality, or the intellectual ability to consider and distinguish fundamental, bipolar concepts (such as "just" and "unjust," "good" and "bad") through speech. Other animals, Aristotle states, have no "perception" of such things, even though they feel "pain and pleasure," as humans do, and are apt to "signal" it to "each other" by using their "voice." That is to say, in humans, mental "perception" is influenced by the faculty of language (in general). "Logos," then, not only refers to language as it is realized through discourse; it also implies that the human mind, in its logic and ordering of the real, depends on *having* language. Because Aristotle is both a philosopher and a naturalist, one could go one step further. *Stricto sensu,* an animal having at least some human language and beginning to perceive concepts through them could consequently display some moral behavior and rearrange its mental life through internal and external dialog *(dia-logos).* Then, this animal would become human "by definition."

From this perspective, the work of Sue Savage-Rumbaugh, who

immersed a few bonobos within human semanticism and gave them the ability to *voice* more than pain and pleasure through the use of a computer and with recourse to arbitrary lexigrams coding English words, created *definitional humans*. This is radical experimental philosophy—the perceptions of concepts such as "good" and "bad," implying a bivalent logic in a social context, are present in these languaged specimens of *Pan paniscus*. The systematic of dialoguing is also remarkable in these animals taking turn, answering (some) questions, and being able to initiate verbal contacts. On the basis of what has been published about these bonobos, on videotapes of them, and on my own direct experience on several occasions, I have no doubt that Kanzi and his family have some access to *logos*. By following the letter of Aristotle's argument, as Savage-Rumbaugh tacitly does, there is no philosophical way to exclude them from mankind—unless one banishes some "sub-humans" from humanity. Granted, these bonobos may not "have" *logos* as I do, but, at least, they forcefully *take part* of it.

The subsequent difficulty is precisely that Aristotle, just after having made in his treatise what appears to be a universal claim about human *phusis* (we are *animals* that *have logos*), puts a sub-category of men and women to the periphery of the group. While acknowledging that legal servitude is above all a social construct, Aristotle also defends this institution and states it is nonetheless justified in the case of people who would be "slaves by nature [*phusei*]." Those are individuals who can only *take part* in *logos* without fully *having* it and who, for their own good, need to be ruled over by free citizens. The colonial and racial consequence of this line of thought continues today to be evident enough. It cannot be easily escaped. Actually, the *definition* of the human is always politically motivated to a certain extent: by designating the in-group, it delimits an ordering of the social project that is put in common between agents. No scientific protocol will save us from this.

SELF-SPECIFICATION

81.

In what ways do *we think of ourselves as human animals*? This is my problem, and it is not bounded by "cognition" only. The *archeological* interventions, from paleontology to psychological speculation or experiment, should be rearranged into an intellective *etiology.* There are two crossing paths the bipedal, cerebrally sophisticated, eusocial, self-domesticated, symbolic, and speaking animal *Homo* must take to claim to be "human": *imaginary recognition* and *performative qualification.* All the other conjectures I briefly evoked rely on this dual requisite.

82.

Animal individuals living in groups are organized by the dynamics and rules of the society in which they take part. By a number of different cues, varying across species, social organisms are able to identify a (presumably variable) degree of similarity with other species. These mechanisms, which do not need to be "conscious," are already crucial in individual reproduction and survival; they acquire a higher necessity in groups. No society would exist without the *a minima* recognition of a *similar other.* An ethics based on the transcendence of the—human—Other, as developed by Emmanuel Levinas, is just the limit-experience of the biological collective. Its particularities lie in a quasi-universal expansion of the similar other, whose contours are always less obvious than one would assume. In positing the recognition of the other *before* the advent of the self, Levinas was certainly right—even though the case for human uniqueness is dubious. What seems to be more specifically attached to "us" is the absolute commandment ("Thou shalt not kill") that the philosopher ascribes to our experience of face-to-face. The similar other is not a priori transcendental: it becomes so through the various modes of social anthropogeny to which we are submitted from the day we were born (and, in a sense, even before this). The verbalized commandment does not only express the status of the other: it creates the Other, on the basis of a mechanism of recognition that is widespread among animals, and it loops back to it, in the course of intellection. This movement already presupposes an ideation of the human that has to be acquired through the image of the self, mediated by otherness.

83.

Having reached a given developmental level, and after exposure to their reflection in the mirror, some animals are apt to grasp that they see themselves. The observation with human infants has been made famous by the early work of Jacques Lacan. Decades before, Charles Darwin mentioned the possibility of testing animals in general, and the *Gestaltist* Wolfgang Köhler conducted some experiments with apes and mirrors in the 1910s. Gordon Gallup, in 1970, designed what became an ordinary protocol to evaluate the ability for "self-recognition" (this involves habituation to a reflecting surface and "marking" individuals with paint or any comparable means to elicit a reaction). It seems accepted that individual elephants, orcas, bottlenose dolphins, magpies, and great apes also "pass" the test. The results may and do vary, according to contexts, which is true of human children too.

For any self-recognition to take place, a series of mental operations is needed, which includes at least physical *perception* (including proprioperception), *discrimination* conferring objectivity to what is seen in the mirror, *information processing, logical categorization,* and (interrupted) *identification* by the agent of her own reflection. Discrimination could be partially encoded genetically and encompasses the recognition of a similar other. At this level, most animals have no need for any further discrimination. They do not grasp that what is perceived is a flat (and odorless) image, and there is little to no difference in the apprehension of a reflection or of a scene from afar. A cat continues to see itself as an intruder and displays signs of hostility. In other species, subjects may discern the limits of the mirrored image, for instance, and deduce that visual cues appearing in front of them refer to what is behind them. Macaques, while unable to recognize themselves in the mirror, have been found to use the reflected image of their environment to locate objects that normally were out of their sight. "Association" could explain many things, but I believe it is via informal logical binding of different observations that an animal individual first understands that the other in the mirror depends on itself. Touching the reflected body is a hint; noticing that the gestures I do are being done "at the same time" by the object is another element. This recognizable image depends on me, while it is at a place that is not my own—then, it could "be" "me." It is difficult to admit that self-awareness is created ex nihilo by the consequence of a reflected image. Indeed, human babies are not competent with mirrors before having a basic mental map of their own physical contours. The identification is, in its own right, surprising. Some combination of bewilderment, amusement, and incredulity is usually expressed in animals reaching that stage for the first time. Despite the natural predominance of bivalent situations, the not-me appearing as a similar other is envisaged as me, through a nonmonotonic revision of my comprehension of the real. Although I was aware of my sensorimotor body before (or aware of *this* body), I now build an additional relation between me and me, that is, *myself.* This sense of the "self" is an ideational *construction of my own otherness,* and its consolidation into a *reflexive* category, that could be detached from me. There used to be some homeostasis, some self-awareness (furnished by proprioperception, it was categorically *selfless*), as well as the circumscription of others, to which I almost spontaneously synchronize or respond. And now, identification bifurcates in the reflexive; then, language could aim to fill in the gap.

The status of the mirrored *me* is far from simple. What I see is me, and it is not. The image of myself appears through this contradiction. This allows a regime of minimal fiction that is often perceptible in playful reactions in front of the mirror: the real has been displaced. The structural possibility for real fiction[57] is also inscribed in this moment, and it is certainly no coincidence that mirrors are ubiquitous in the fantastic or that reflexivity is a noted trait of the literary. As a matter of fact, children's plays in front of mirrors (a behavior observed in some humanly encultured apes as well, and possibly in some dolphins) signal that the reflecting surface is integrated into the cognitive apparatus. In more complex situations, the mirror serves as a Φ-operator, authorizing a

contradiction, whose signification is lived through experimentation and the assemblage of myself. The derived *self*—always the
same, always the other—is conceptualized through the imaginary
constitution of something that is both *like you* and *like me*.

This is where another *semantic* step is reached. By association,
the image has the sense of another *I*—and, by dissociation, the
sense of *another* me. The reflection becomes *meaningful.* Hence-
forth, the imagination of the self is tied to a semantic extrapolation
where the abstracted reflection of a conspecific virtually contains
me as well as others. In an etiological stance, which, once again,
does not necessarily coincide with the natural archeology of the
phenomenon, self-recognition in a reflection prepares the notion of
the human, as an *imaginary repository for all virtual selves.*

84.

The humanness of the human is additionally produced in a performative speech-act. As with all discursive events, the proclamation
of the human is usable outside of its context of enunciation; it also
has to be reiterated to stay consequential. As soon as an animal
is able to qualify itself as *another* kind of living object, a conceptual existence is advanced and could be refined, modified, or pre-
scribed. The limits of the human, here, are no scientific *datum.* The
image of the other human is structured by perlocutory discourse.

In many languages, the expression of the reflexive happens
through the special lexicalization of a noun referring to the body
(in total, or in part, from *bone* and *heart* to *soul* and *skin*).[58] Other
known "strategies" for saying the reflexive recycle words for "alone"
or "only," or for "the same." The demonstrative pair **swe/se* from
Proto-Indo-European has been used to construct the third person
(German *sich,* Latin *suus,* etc.) as well as the reflexive *self* and its
equivalent forms in other languages. All in all, discursive categori-
cal reflexivity is generally uttered through a primordial reference
to the reflected *body* image that is the *same* as others and contem-
plated on a *one-to-one* basis.

Thence, a discursive plane could be built, where several selves
are temporarily combined. It seems that most (if not all) names
that human groups give to themselves fall into one of these three

categories: they may be descriptively geographic or etymologically refer to "the free ones" or "the humans." Freedom is acquired in contrast with human slaves but also, and plausibly above all, as a label for groups of animals thinking they have been separated from the rest of *phusis*. As for the redundant designation ("we are the human beings"), it implies both that other human nations may be "less equal" and that what Cornelius Castoriadis called "the imaginary institution of society" is based on a repeatable self-qualification of the human image in a human language. The *we* that we utter is never a simple addition of individuals. It virtually creates a community, where differences are sustained by both the common and the shared. Philosophical descriptions, religious prescriptions, social commentaries, political evocations, legal norms, or scientific reveries are all expressed through and within a sometimes implicit, and often explicit, *we*—all of which concurrently shape "mankind." Language, from its lexical and syntactic mechanisms to the worlds of words it supports, is deeply engaged in the performance of the human. A consequence of mirrored recognition, the "ourselves" of which we speak, is incompletely contained within the discourse that puts it forward. As for the Socratic "voiceless dialogue of the soul with itself," it already signals the intellective displacement of reflection.

85.

Humans are animals that specify themselves through reflexive imagination and speech.

86.

There is a "human nature" in the sense that we, as living animals, are bounded by biophysical constraints. Though this fact has often been occulted, it is nevertheless inescapable. Now, this "human nature" does not tell us that *we* are monogamous, or peaceful, or vegan. This would be a confusion, not between what "is" and what "ought to be" but between the human and the human. A sharp separation between our nature and our culture(s) will not do either. Our cultures are natural through and through, for at least three reasons. Because what is, to us, the inconstructible of *phusis* has well been fabricated to some extent, something the classical dichotomy of nature and culture always disregards. Because the human systems of symbols are still material and naturalized in our brains. Because the imaginary and discursive projection of humanness actually modifies our "biology." None of this is acknowledged, or even understood, by the *phrase* of human nature, which is why we should rather abandon it, once and for all. Yet there is no absolute autonomy or omnipotence of culture to seek.

87.

As self-domesticated animals ourselves, we have benefitted from the relative relaxation of some evolutionary constraints. The gradual extendedness of our minds, granted by inert and animal tools, verbal language, writing, "information technologies," and the like, has expanded our epistemic reach. With supports more durable than the individual brain, we have also expanded the cumulative transfer of skills and expertise through partial conservation of knowledge. This logically came with a necessary improvement of our capacity to handle the deviations of intellection (which multiply in proportion to the performance of cognition through extendedness). The subsequent *extension* was susceptible to mean more, because of its operative lack.

This is the strong position that I defend in this book. New knowledge could be acquired through an automatic process of trial and error; it could be obtained by filling in the blanks of a theoretical frame. Most of this could be interpreted as cognitive output. At best, controlled thoughts of this sort will be tabular, cohesive (or as self-consistent as possible), and somehow open-ended. We need those, and there is no point to advocate a mythic return to the prescientific. *But* the promotion of the cognitive as the only (or even most respectable) regime of human thinking entraps us in a world devoid of anything other than massive reproduction and episodically derived inventions. A much more intense mode of creation is found in the rearrangement of the failures given by rational protocols. The intellective is a surplus stemming from defection: *we have to do the work of cognition as much as we can, before reaching the points of rupture and, from there, thinking beyond.* This kind of "diversity of behavior" is *not* the issue of "storage capacity"[59] Turing neglectfully mentioned in one of his articles.

The machine, in itself, may be a model of what we are. Gilbert Simondon notoriously developed this line of thought in regard to the living and gave a more complex conceptual weight to a tradition particularly marked by the materialist Cartesianism of La Mettrie. The problem lies with the dominant view of the machine as perfect or divine. Under this assumption, intellection should be reduced or, possibly, attenuated. The impassible God of absolute mental operations represents, in fact, the uncreative genre of *noēsis*. A computer fails or succeeds in its cognition, and that is it. Only robots that would be both extremely sophisticated and defective could *create*. Self-organization in artificial intelligence, with robots left to constitute their programs, is a fascinating endeavor. It remains to be proven that, even for these "autotelic agents," as Luc Steels calls them,[60] the conditions of epistemic and noetic consolidation would be comparable to the extraordinary insight we gain from the broken course of intellection.

Becoming a cyborg—the "hybrid of machine and organism"[61] Donna Haraway once celebrated—could mean that a human body is "enhanced" by the addition of electronic chips augmenting computation. Then we would simply be speaking of a particular case of extendedness, as long as the cybernetic machine is still conceived in a traditional way. Mental implants would possibly *minimize* the intellective, and/or reinforce the urge of cognitive reversion, making us highly well-informed and dull animals, with the idiot savant as our ideal. I understand that Haraway also exploits the fictional side of the cyborg in her "manifesto." But I am saying that if we keep the dominant (and *cognitive*) view of the mechanical mind, the cyborg is nothing other than the portable "goddess."[62]

In the same vein, I am impressed when a program (such as David Cope's "Emmy," also known as "Emily Howell") composes *like* Johan Sebastian Bach; but I prefer the unheard piercing through the known. There is little doubt that creation is not ex nihilo and that a process of recombination is at work—database and storage, if you want—but the challenge here is to endow machines with the ability to open up new musical, conceptual, or poetic languages and not *apply* compositional solutions already designed by human minds. This, I say, will be possible *only* if one relinquishes the engineer's religion of the divine machine. We could still habituate ourselves to computer music or poems, corresponding to what third-rank human artists produce, and be happy with it; mediocrity appeals to many. This will not change the fact that *poiēsis* in the sciences and the arts, and not just invention, is the product of the intellective we have both to assimilate through cognition and to revive through our own access to the additional space.[63]

88.

Because we also specify ourselves, the current understanding of our animalness, if we wish to incorporate it into our discursive imaginary, leaves several options. We may foster a *generalized anthropogeny*. This is the common goal of otherwise heterogeneous efforts, such as the political movement behind the recognition of ape personhood, the technophile fascination with the Turing test, and the updated fantasy of an intelligent universe, saturated with human thought.

Another temptation consists in *dehumanizing the human*. The idea has ramifications in historicism, which childishly concludes that because the category of "the human" is constructed, variable, and discontinuous, it should be discarded or reputed to be "ideological." (This is a blatant way of missing the whole performative aspect of the imaginary and verbalization of the self in the collective.) A theoretical emphasis on the transindividual processes of life also went at odds with the conventional image of a self-ruled and conscious subject, precipitating the idea that, all the "human" attributes being suspended, humanness should disappear from our discussions. This consideration, combined with an attachment to "the animal," or (at least) some of them, gave way to the contemporary description of the posthuman, which sometimes rejoins the celebration of the cyborg. The work of Donna Haraway, once again, is exemplary of such a trend.

As for me, I confess that I am not terribly interested in defining the limits of the (post)human. I am as disgusted by humanity as unemotional with computers—and I do not even "own" pets. I do not believe that I can do without qualifying our species at some point, but my aim is rather to focus on the *intellective animal* (be it "human" or not), the one creating and sharing signification.

89.

The call for "posthumanities" to come is both comprehensible and legitimate. The institutional architecture of the humanities is crumbling, and its social value is shattered. Humanistic universalism is suspicious to most, because of what it politically recovered and the current changes in anthropogeny. The defense of the humanities as "what makes us human" has largely become inaudible. The centering on *tradition* (instead of transmission) à la Hans Georg Gadamer is lethal—just another version of historicism.

Would the posthumanities no longer be "anthropocentric" in any respect? I doubt it, inasmuch as their expression stays within the boundaries of verbal language, of knowledge, and the imaginary escorting them. A forceful recourse to the deus ex machina would be pointless, because a truncated version of the human mind is precisely encrypted in the perfect computer. Willy-nilly, we are going to retain something of the human *animal* as long as we think together; and "centers" or zones of relative stability will still emerge in *dianoēsis*. Thus, the issue is rather to reconsider both the *anthrō-pos* and the *kentron* so significantly that we would contribute to the fostering of our self-specification, without falling into the mere repetition of religiously inflected anthropocentrism.

The emphasis on the intellective is a theoretical, epistemological, and political project. It cannot accommodate the present state of globalized academia, where poetic thinking is being marginalized at best and censored or ridiculed otherwise. Graduate studies, in particular, are a factory, where early careers are being mainly decided on an ability to obey and to conform, and where the desire for the unexpected stemming from organized knowledge is violently repressed. This paves the way for the entrepreneurial bureaucracy of the experimental (or the applied) in what Thomas Kuhn dubbed "normal science" and, in the humanities, for the interplay between the empty stutter of professionals mutually glossing each other in a grotesque quest for authority—and the preepistemic subjectivism of "my identity."

The most discursive disciplines, if they want to express something significant and displace the institutional body of the humanities, have to speak of, and from, the intellective, in a manner the regular sciences could contribute to but not appropriate.

90.

In a flash of light, forms appear on the stone. The auroch has been painted on a curved rock, so that its two-dimensional picture could gain some depth. A series of dots underline the image of a bison and have been added afterward. The head of a small manly figure has the beak of a bird. The prehistoric cave paintings illustrate the commonness of *Homo sapiens* in an apprehension of the real and fears or emotions that are still perceptible to us, even when detached from the culture and legends in which they once partook. Painting is a cognitive act, whose time and durability are biological. Also, the loss of the special *sense* that may have been encoded in these drawings and paintings cannot annul the signification of these representations for us. Meaning is sustained by the stability of the cognitive apparatus, but it is created anew from the intellective space. And there, obviously, what we see of "ourselves" through the distorted mirror of prehistoric artistry are our own images, thousands of semantic signs (dots, arrows, triangles, etc.) and animal figures, sometimes including a few apes like us. We are the animals . . .

91.

The human animals are mortal. This is another defect of theirs. No ruse, no machine seems able to cure this ill.

92.

Stories and images abound of immortality, afterlife, and resurrection. Plenty of them imply some post mortem survival of thought. It is tempting to identify most beliefs about a perennial "soul" of ours with a mystical apprehension of our mental capacities. The neopositivist stance would argue that the central nervous system is the source of all our illusions regarding the preservation of mental abilities. It is equally easy to respond that, as useful as the brain is for our "minds" to appear, we have no perfect correlation so far between our own impressions and their neurological support. Our models are very crude indeed, and there is a lot to discover. As I said, there are sound reasons to estimate that our science will never be complete. I do not try to prevent anyone from believing in the sheer assurance of human knowledge, or in collective and individual revelations; such is not my purpose. The third, "religious" question Immanuel Kant saw as a pillar of "the field of philosophy," that is, "what may I hope?"[64] has a straightforward answer: "everything I can." There is no conceptual route guiding me, once and for all, from the sea of errors to the tiny pond of rational hopes. My visions of God and mortality are constrained by what I can picture and describe, no doubt, but they do not have to be bound by logical computations, or even ideals of coherence. I do not need to be and act in accordance with myself. After all, I am not completely sure of what I believe in—many small reactions of mine show like a discordance within me. I may be influenced by arguments or shocking experiences in such a way that I will not promptly "change my mind" despite other pressures. In this regard, a long time spent studying brain imagery or a "near-death experience" would have comparable effects, and, in terms of *hope,* one attitude is frankly worth the other. The Kantian naïveté, anchored in the traditional precepts of philosophy, is to connect rational discourses and hopes in a unidirectional fashion.

93.

First preliminary. We should suppose neither linear progression nor pure immobility in the human descriptions of death and souls. On the developmental side, the conception of Piagetian stages is mainly outdated. To propose supernatural explanations of the unknown, some maturity is required. Superstitions are not the established foolishness of early childhood. Adults, in all societies, are the ones building religions. Beliefs about spirits are apt to coexist with accurate and "medical" descriptions of corporal death and the dissolution of the brain. On a broader scale, human populations do not "progress." A sequence à la Frazer "animism giving polytheism giving monotheism giving truth" is not an insane babble: these religious inventions are indeed interlinked, and one stems from the other. But the sequence, as such, is flawed (even without its final step): it conceals the persistence of each "previous" stage as well as its partial blending into a new doctrine; then it masks all the other subproducts to the benefit of a fallacious historical *order*. In such a context, the teachings about the nonmortal parts of myself vary in a nonlinear manner.

94.

Second preliminary. In most civilizations, the vocabulary and imagery of the psychic and the spiritual are particularly rich, complex, and overlapping. It is often possible to find cultural strata; the Khmers, for instance, could have recourse to the Buddhist *viññāṇ*, to the nineteen "pagan" *braliṅ* that each individual carries with herself, and to the Christian soul.[65] One may explain aspects of provenance. However, these mutually incompatible systems work together. Furthermore, if taken "one by one," they are already traversed by incoherence and are discontinuous. At any rate, a special conceptual mobility for the "souls" is the effect of the both fleeting and evanescent object it tries to characterize and of the blind spots of metacognition.

95.

We have the mind. We are not at ease if we want to define it—though this situation is far from unique to us—but the mind we can figure out is *embodied*, that is, cerebral and more than this; also *extended* to other discrete material supports; and *emitting signals* of sorts when active. The range of consciousness among animal minds is quite variable, with a good deal (perhaps most) of mental activities happening below the threshold of the conscious. The normal quality of our mental life out of sleep may be due to the evolutionary peculiarities of primates, whose circadian rhythm seems to foster an "adaptive waking behavior."[66] It seems hard to deny this kind of consciousness to all the nonhuman animals—unless one asserts that only verbal consciousness (or mind) exists, or that the conscious is the ability to self-report with words. To me, this would be conflating a given range of the action of mind (consciousness) with at least one of its products (the discursive consolidation of the self), and possibly with the semantic and syntactic construction of personal agency through language (the *I* that I am). The "first person" is an index of ideation that shapes me through the process of intellection. If we want to think of ourselves as animals and stay with defectiveness, we ought to avoid exactly *equating* verbal construction with psychogenesis. Being more or less conscious, *memory* finally sustains a temporal cohesion of the mind and the self, by selecting and rearranging encrypted moments of the past. It also fills the episodic *I* that will be further developed and altered, before being possibly encoded into a changing self—and before restructuring mental activities, and so on.

In this schema, individual afterlife is often seen as a conservation of the mind holding on to its plastically recorded and retrievable memory, its fabricated self, through first-person expression. The mind would be disembodied and thus freed from its earthly existence, or "teleported" into another material vessel, such as an "astral form," a computer, or a resurrected body. From this perspective, what is called the mind by the cognitive sciences and others is largely coextensive to this religious *soul*. As a result, in the scientific approach as it currently stands, there is no soul but the mind, and if there is no way to "store" it appropriately after death, it simply vanishes with the body. It may be, however, that a soul has nothing to do with the mind or is so fundamental to its emergence and functions that it could not be registered by current instruments or theories, in which scenario, almost nothing of what we discussed so far actually bears on this topic.

Other mystical doctrines could refer to nonmaximal "souls," or disembodied minds that would operate only unconsciously, and/ or losing memory, suspending their selves, and/or bypassing the verbalized and perceptible *I*. The ravishment of the souls, their transmigration, their lethargy, are all possibilities considered by traditional religions or "new age" myths. It seems that the most recent and pervasive monotheistic religions (Christianity and Islam) usually have a thirst for maximal depictions of the posthumous soul (as the self-conscious, hypermnesiac mind speaking to God), which today translates into technophile evocations of the eternal human-machine. Anyhow, we are left with the choice between the conceptual assimilation of mind and soul, implying the open question of material conservation for some postponed technological future—and the assertion of a superfundamental or absolutely autonomous (and so far imperceptible) soul, deferring this discussion to the day of revelation, where all contradictory conversations will be made de facto futile.

96.

The *psukhē* of psychoanalysis is a certain construct of the mental, where the moving line of the conscious is endowed with very high meaning. In accordance with Freud's scientism and his first project of "scientific psychology," some scholars are currently trying to bridge neuroscience with at least some psychoanalytical concepts. This endeavor is just on its way, and I cannot picture how it would continue without proceeding to some major revisions of the original dogma(s).

The psychoanalytic "soul," as it is currently described, lies in the interaction of the Ego, the Id, and the Superego. Their ligature commands an economy of affects and drives the production of thought. Then, conceptually, this "psyche" adapts the philosophical Greek *psukhē,* under the human conditions of mortality and with an insistence on its own, opaque, divisions. It remains that a key purpose of the discursive cure, and a feature of the Freudian doctrine itself, consists in making the nonconscious conscious. In articulation with its relatively fixed "topic," the psychoanalytic process is also moving (along) the line of consciousness. In this model, the mind becomes a subcase of the soul or one of its manifestations. Conversely, this *psukhē* serves as a general mind-set, as a frame of reference. Memory is essential as a repository of scenes that are reinvented according to apparently universal rules or fantasized in accordance with the same laws, with a putative influence on all behavior, coming from the background. Thus, it seems that the psychoanalytic soul is usually conceived as the individual expression of the mental, shaped by "natural" and stable parameters (the biological function of sexual reproduction, the need for a body image, etc.), general human cognitive abilities (language, narration, symbols, etc.), particular development in a social environment— and governed by unconscious affective "drives."

All in all, this portrait of the "soul" differs from the cognitive mind in its prerequisites and conclusions, but in fact, it is not exactly something *else*: we just have large discrepancies in methods and beliefs, leading to slightly or largely incompatible constructions of the object, without conferring actual singularity (or even detachability) to *psukhē* in regard to *nous.* Accordingly, the name of *psukhē* is usually kept in the etymological rubric of *psycho-*analysis but is rarely further quoted, or only to be *exposed,* to the benefit of "the subject" and its topical polarities. Even the Jungian distinction, borrowed from the Latin, makes *animus* and *anima* the two sides of the same coin. If we try to create a new discourse of the soul piercing through the mind, psychoanalysis is of little support.

97.

One could see the *qualia* as the philosophical name for entrenched soul. The mental "what it is like" to feel, speak, or imagine has been largely appealed to, in Anglophone philosophy especially, each time proponents of "strong artificial intelligence" make their case. It could turn out, as Stanislas Dehaene argues, that these unique *like* perceptions do in fact transcribe the *access* to consciousness, in which case they would not be enough to disqualify cognitivism. But Dehaene's explanation has no bearing on the more general argument (first introduced by Lewis in 1928, before the modern theories of computer). *Qualia* could still be mentioned as what sustains the qualitative and nontransferable aspect of consciousness, or as an *impression* that could be recorded by an fMRI machine or a scientist but not felt by them. This, I believe, is the contemporary manner to enclose the soul within the mind, almost to entrap one into the other: through the concept of *qualia,* the soul becomes the innermost consciousness.

Even if the word *soul* is not employed by analytic philosophers when they refer to "zombies," their representation of human body doubles lacking *qualia* is undoubtedly a picture of our soulless selves. In Haiti, zombies are supposed to be people held between life and death, who are manipulated and exploited by voodoo priests. The "philosophical" version, also influenced by (post)colonial Hollywood B movies, is clearly disturbed (deranged?) by issues it rarely seeks (or succeeds) to problematize: the relations between soul and mind, the survival of anything material or immaterial posterior to the decay of the animal body, and so on. Apostles of universal rationalism are certainly caught off guard, as they refer to fantastic zombies under the guise of "thought experiments."

98.

In many civilizations, something like "the soul" is what *moves the living.* Despite many doctrinal rewritings, it is worth repeating that, for Aristotle, all that lives (even plants) has a soul. As for Plato, in his work, animals are clearly endowed with *psukhē.* Indian Jainism or many other traditional religions, from Africa and elsewhere, suppose a kind of force *animating* natural entities. In other words, the restriction of "the soul" to the human realm is just one position, even within the (post)Greek tradition.

In its generalized view, the soul is another possible name for what now appears to be the self-organization of inert matter into a living organism. More precisely, the language of "the-soul" attributes a meaning to senseless emergence and contains ad hoc physical processes inside a metaphysical matrix. In the "animal" hypothesis, the soul is either a restricted copy of the living principle we just evoked or a *fundamental product* of individual experience. The redistribution of the souls after the death of the body, through a cycle of migrations depending on the deeds and misdeeds of each individual over the course of its terrestrial life, illustrates the fact that the transcendental principle of "animation" is both fundamental and derived in content from its previous incarnated existences. Here the morality of the soul-bearer plays a heavy role. Then, to say it plainly, were the soul a semantic feature designing an arch-principle of *phusis* or animality, we'd have little to add. And if by soul we mean the ineffable "qualia" of mind itself, something being modified by what I feel and think, we are back to a previous problem. In some respects, one could say that the conception inherited from Plato includes the mind into the soul, whereas the contemporary tendency would consist in reducing the latter to the former—as a way to preserve both or in an attempt at getting rid of anything psychic.

99.

The *anima mundi,* as a mindless force, ineffable and devoid of self-agency or consciousness, corresponds to a theoretical separation of categories. But, on one hand, the reflexive feeling of what I am will still perish with my body after death. On the other hand, this large Soul looks a lot like the *Nous* of cosmos, without direct connection to the content of our thoughts.

100.

After having listened to Socrates speaking a few hours before drinking hemlock, Simmias (who is one of the disciples gathered around the master) makes the following objection: the whole demonstration that has been developed about the immortality of the soul could be applied, but to no avail, to "some lyre and its chords."[67] Conversely, it might seem more reasonable to admit that, in the same way musical harmony is lost when the lyre is broken, "in what one calls death, the soul is the first to be destroyed."[68] Socrates ironically lauds his young friend for his independent spirit and even feigns to be in trouble with his own argument. Then, he resorts to his usual strategy and shows that his interlocutor had previously accepted several other points in the course of the dialogue—and that it would be inconsistent to go from them to the analogy with the lyre.[69] The proof is enmeshed in a standard (and semiformal) logic that is supposed to "give reason" of the real. What are the arguments Simmias conceded? That, from the body, "no thought ever came to us, never";[70] that a bivalent and dialectic logical structure commands what is, including the fact that life proceeds from death, and vice versa;[71] that innate thoughts are established through the advent of reminiscence, suggesting that something tied to our soul preexists human birth and the constitution of the body; and that, the noncomposite being impossible to dissolve, the soul, which is "almost a whole,"[72] is consequently without parts. True, after having admitted these doctrinal elements, the comparison with the harmony created by a musical instrument is much more ungrounded. But what happens to this idea, if we believe that the body thinks, that logic does not transcend *phusis*, and that mental inborn aptitudes and conceptions are the product of biological evolution?

101.

Here is what I want to suggest. We may doubt of a cosmic soul or admit that most attributes of *psukhē* (if not all) are now understandable as embodied synaptic emergence. But we may aim to keep the categories of the soul. As for the eventual survival of our *spirit,* disembodied, floating through the air, transferrable to a new living site, I gladly leave you judge of it, and of what you decide to believe. Turing said en passant in one of his most famous articles that all for which he was arguing could be ruined by the recent discoveries of parapsychic activities, if they were confirmed (they were not). So I can certainly allow myself a possible exit toward the paranormal. After all, I'd love to be wrong about the mortality of my mind and spirit. In the opposite case, which I presuppose here,

I may still speak of my soul as an effect of this mind toward another I than the one I think I am using.

My soul is a singular persistence of my thinking into yours, yours, and yours.

102.

A shortness of soul is obvious in all informative utterances and referential actions. Through the extension of intellection, something more than my mind (and less than itself) is apt to appear. The soul is manifested on the intellective space. It is affected to a self through the movement toward cognition: after I witness the shared soul, it becomes mine and flees. Our souls are not enchained to ourselves, to our consciousness, to our mental processes, to our memory, even though they can relate to them. Reading literature, being immersed in art or love may be moments of ecstasy, of rapture, of possession, with me falling into a detachable soul that is uncommon though already shared.

103.

On the verge of science fiction, a recently formed foundation[73] advances the idea that, with the appropriate progress in neuromapping and some further development in the chemical or cryonic preservation of tissues, the content and idiosyncratic modus operandi of a human brain could be saved, right after death—and possibly resurrected, re-created on another material support, be it another "wet" nervous system or a computer. So far, absolutely nothing guarantees that the integrity of a complex and fragile organ like a brain could be effectively preserved, by any known or foreseeable technique, and that some neglected "minor details" would not ruin the entire "conservation effort." Furthermore, establishing the connectome of an individual is, to this date, something that has not been accomplished; actually, as we said before, we do not even have a model for an animal other than a nematode worm. I do not speak of the difficulties posed by the artificial replication of neurons, synapses, and mental routes. I do not recall the numerous (and quite justified) reservations that have been raised against "strong artificial intelligence." In sum, we might be *(very)* skeptical, but, given what we believe to know of our brains, I cannot find a good reason to reject this wild fantasy en bloc. Something could be saved. In fact, we are told by the foundation that portions of the past could be kept, such as memories. "Full revival" or "life restoration" is, nevertheless, the goal. A *full* revival is extremely unlikely, and all the more so in the decades to come (for this is the time frame mentioned on the website). Then, to use appropriately the expression "life restoration," we would need to identify animal life with brain activity, to assume that "living" *in a vat* is livable, and/or to create new robots or chimeras as bodily "hosts" for our cognition.

Now, let us suppose that we can engineer such a *transmigration of the minds.* How are we going to think? Randomly, because all of our beautiful theories of the mind are profoundly wrong? Partially, because the structural incompleteness of our knowledge and our persistent ignorance would lead to truncated replications of ourselves? Imperfectly, because we lack the techniques to make things work as they should and thus would have to painfully witness the revival of a brilliant mind as an incapacitated intellect? Perfectly but repeatedly, because the new brain would exactly reproduce the connections it has been programmed to redo? Or, because, as a major advocate of the preservation project asserts, we are so "repetitive"[74] anyway that even a crude digital simulation of ourselves is able to guess what we want "most of the time"?

But what I want is to think *defectively,* that is, to be able to operate at a maximum *and* to be displaced by intellection. Reviving my mind in a purely cognitive mode would be a way to catch and destroy my very soul.

104.

Our souls are "immortal," for they do not *live.* They have a *translife,* they are differentially performed by organisms. They disappear, when painted stones are erased, languages go extinct, memories vanish. But, as long as their *tracé* is to be found, they will appear, and sink.

105.

Dialogic of the soul: it is me, it is not. *Extraordinariness* of the shared and transient "very high order" of thinking. *Defection,* already, rapidly. *Affirmation* nevertheless of a beyond that is not, or no longer— but is still performed.

106.

Through the intellective, I augment the possibilities of my soul, and I expand it, up to the point where something will virtually stay from it, independently from my perishable brain. This is not the resurrection of bodies, a mythic promise that is still being formulated today. This is the best way to persist beyond ourselves, despite the defectiveness of our real lives.

107.

As the lyre will be broken, I will no longer be there to relate my self to the singular soul of which my mind once dreamed. I surely hope you will. ◊

Acknowledgments

This book would not have been possible without the generous support of the Andrew Mellon Foundation, the award of a New Directions fellowship (2009–12), and a supplemental grant (2012–15). Thanks to the Mellon Foundation, I was able finally to find the time to properly study areas of research I had wanted to explore in more depth since 2002–4 and the editorial work I was then doing for the journal *Labyrinthe*. This led to the current book and to many other developments, including my affiliation to the cognitive science program of my university—a move that is still slightly eerie to me.

I had some decisive discussions about several aspects of this book with Marc Aymes; Jim Benson and Bill Greaves and their undergraduate researchers at Glendon College; Sacha Bourgeois-Gironde; Morten Christiansen; Stanislas Dehaene; Jean-Louis Dessalles; Alex De Voogt; Amable Dufatanye; Paul Égré; Laurent Ferri; Bill Fields and his employees at the Great Ape Trust; Bruno Galantucci; Tristan Garcia; Robert Goldstone; Uri Hasson and his colleagues at Princeton University; Peggy Kamuf; Anthony Mangeon; Renaud Pasquier; Pietro Pucci; Liz Pugh; Sue Savage-Rumbaugh; David Schreiber; Benjamin Spector; Luc Steels and his collaborators at the Sony Lab in Paris; Karyl Swartz; Ian Tattersall; Ioana Vartolomei; the Wamba family; Spencer Wells; and Cary Wolfe. I am particularly grateful to all the scientists who welcomed the outsider I was.

I want to warmly thank the Cornell University graduate students who attended my 2012 seminar on "Literary Theory and Cognitive Science": David Aichenbaum, Neal Allar, Jordan DeLong, Evan Foster, George Karalis, Matthew Kibbee, and Jacob Krell. I was subsequently invited by Anthony Mangeon to present a much condensed version of this seminar when I taught in May 2012 as a visiting professor at the Université de Montpellier, and I also express my gratitude to all the participants there. Between 2010 and 2013, I publicly presented some aspects of this research at Cornell University, Princeton University, Simpson College, the École des hautes études en sciences sociales, the École normale supérieure in Paris, the École nationale des Mines in Saint-Étienne, and New York University. I am indebted to all those who made these discussions possible.

Finally, I have received the most crucial help from the first readers of my book manuscript: Douglas Armato, Marc Aymes, Laurent Ferri, Matthew Kibbee, Brian Lennon, Cary Wolfe, and an anonymous reviewer.

Notes

I. THE INTELLECTIVE HYPOTHESIS

1. Throughout the book, footnotes are mainly used to provide references for direct quotations. For additional content and observations, see the "Repertory" at the end of the volume.

2. Claude E. Shannon and Warren Weaver, *The Mathematical Theory of Communication* (Urbana: University of Illinois Press, 1949), 5.

3. Ibid., 3, for all the quotes appearing in this sentence.

4. Shannon, "The Bandwagon," *IRE Transactions: On Information Theory* 2-1 (1956): 3: "Information theory has, in the last few years, become something of a bandwagon. . . . It is certainly no panacea for the communication engineer, or, *a fortiori*, for anyone else."

5. René Descartes, *Meditationes de prima philosophia*, II, 5.

6. Edmund Husserl, *Cartesian Meditations: An Introduction to Phenomenology*, § 14–22.

7. Gilles Deleuze, *Différence et répétition* (Paris: Presses universitaires de France, 1968), 183–84.

8. Gerald Edelman, *The Remembered Present: A Biological Theory of Consciousness* (New York: Basic Books, 1989), 64–65 in particular.

9. Baruch Spinoza, *Ethica* III, "affectum generalis definitio" *(in fine)*; my transl.

10. Gregory Bateson, *Steps to an Ecology of Mind: Collected Essays in Anthropology, Psychiatry, Evolution, and Epistemology* (New York: Ballantine, 1972), 410.

11. Cf. Jean-Yves Girard, *The Blind Spot: Lectures on Logic* (Zurich: European Mathematical Society, 2011), 19–20.

12. Karl Popper, *Objective Knowledge: An Evolutionary Approach* (Oxford: Clarendon Press, 1972), 106.

13. Ibid., 107–8, for instance.

14. Greg J. Stephens, Lauren J. Silbert, and Uri Hasson, "Speaker–Listener Neural Coupling Underlies Successful Communication," *Proceedings of the National Academy of Sciences of the United States of America*, 107-32 (2010): 14425–30.

15. Plato, *Sophist*, 263e; my transl.

16. Maurice Merleau-Ponty, *Signes* (Paris: Gallimard, 1960), 21 ("si

elle [la pensée] se maintient, c'est à travers"). The translations in *Signs* (Evanston, Ill.: Northwestern University Press, 1964), 14, and Leonard Lawlor and Ted Toadvine, eds., *The Merleau-Ponty Reader* (Evanston, Ill.: Northwestern University Press, 2007), 329, both rest on a misreading of the original.

17. Based on Quian Quiroga et al., "Invariant Visual Representation by Single Neurons in the Human Brain," *Nature* 435-7045 (2005): 1102–7.

18. See Sebastian Seung, *Connectome: How the Brain's Wiring Makes Us Who We Are* (Boston: Houghton Mifflin Harcourt, 2012), xv.

19. Catherine Malabou, *Plasticity at the Dusk of Writing: Dialectic, Destruction, Deconstruction* (New York: Columbia University Press, 2010), 82.

20. Julien Offray de La Mettrie, *Oeuvres philosophiques* (Berlin, 1774), vol. 1, 113; my transl.

21. La Mettrie, *Oeuvres philosophiques,* vol. 1, 154; my transl.

22. Cf. Malabou, *What Should We Do with Our Brain?* (New York: Fordham University Press, 2008), 72–81.

23. Andy Clark and David Chalmers, "The Extended Mind," *Analysis* 58-1 (1998), 7–19; and, more generally, Richard Menary, ed., *The Extended Mind.* Cf. also Gregory Bateson, *Steps to an Ecology of Mind,* 454, and Merlin Donald, *Origins of the Modern Mind: Three Stages in the Evolution of Culture and Cognition.*

24. Drew Rendall, Michael Owren, and Michael Ryan, "What Do Animal Signals Mean?" *Animal Behaviour* 78-2 (2009): 233–40.

25. Michael Tomasello, *Origins of Human Communication* (Cambridge, Mass.: MIT Press, 2008), 114.

26. Ibid., 254–55.

27. Ibid. 254.

28. Ibid. 255.

29. Walt Whitman, "Song of Myself," stanza 51, *Leaves of Grass.*

30. Gertrude Stein used multiple versions of this phrase throughout her work, including "Rose is a rose is a rose." Cf. her own comments in *Lectures in America* (New York: Random House, 1935), 177: "Each time that I said . . . that somebody was something, each time there was a difference just a difference enough so that it could go on and be a present something."

31. Noam Chomsky, "Minimalist Inquiries: The Framework," in *Step by Step*, Roger Martin, David Michaels, and Juan Uriagereka, eds. (Cambridge, Mass.: MIT Press, 2000), 92. Cf. Chomsky's 2007 talk "The Biology of Language Faculty," https://techtv.mit.edu/videos/16291 -the-biology-of-the-language-faculty-its-perfection-past-and-future.

32. Chomsky, "Biology of Language Faculty."

33. Paul Churchland, *Neurophilosophy at Work* (New York: Cambridge University Press, 2007), 159.

34. Blaise Pascal, *Pensées*, Brunschvicg ed., fragment 1; my transl.

35. Martin Heidegger, *Was heisst denken?*, in *Gesamtausgabe* (Frankfurt: Klostermann, 2002), vol. 8, 9; my transl.

36. Ibid., 41; my transl. Cf. Heidegger, *Sein und Zeit*, § 3.

37. Heidegger, *Was heisst denken?*, 109; cf. the end of *Was ist das—die Philosophie?*

38. Heidegger, *Was heisst denken?*, 9; my transl.

39. René Thom, *Stabilité structurelle et morphogenèse* (Paris: Inter-Éditions, 1977), §13.1.A, 294–98, my transl.; *Modèles mathématiques de la morphogenèse* (Paris: Union générale d'édition, 1974), 114; *Morphogenèse et imaginaire* (Paris: Circé, 1978), 47–48 (about "static" and "metabolic fusion").

40. Michael Beaney, ed., *The Frege Reader* (Malden, Mass.: Blackwell, 1997), 47–52, 149–80.

41. Alfred Tarski, *Logic, Semantics, Metamathematics* (Indianapolis, Ind.: Hackett, 1983), 165.

42. Ibid., 153.

43. Rudolf Carnap, *Logische Syntax der Sprache*, § 1–2.

44. Ronnie Cann, *Formal Semantics: An Introduction* (New York: Cambridge University Press, 1993), 1.

45. Aristotle, *Metaphysics*, Γ, 3, 1005b; my transl.

46. Ibid., 1005b; my transl.

47. Ludwig Wittgenstein, *Tractatus logico-philosophicus*, § 7, "Whereof one cannot speak, thereof one must be silent."

48. Ludwig Wittgenstein, *Remarks on the Foundations of Mathematics* (Cambridge, Mass.: MIT Press, 1978), 110 (original German), 110e (English transl.). I am correcting the English wording, which erroneously gives "a quite different light" and attenuates Wittgenstein's point.

49. Heraclitus in Hermann Diels and Walther Kranz, eds., *Die Fragmente der Vorsokratiker* (Berlin: Weidmann, 1952), fragment B 50; my transl. for all the fragments quoted in this paragraph.

50. Ibid., B 62.

51. Ibid., B 51.

52. Ibid., B 103.

53. Ibid., B 60.

54. Ibid., B 123.

55. Ibid., B 50.

56. Ibid., B 93.

57. Lucien Lévy-Bruhl, *La mentalité primitive* (Oxford: Clarendon Press, 1931), 21–22; see also *Les fonctions mentales dans les sociétés inférieures* (Paris: Presses universitaires de France, 1951), 112–13.

58. Lévy-Bruhl, *La mentalité primitive*, 21; my transl.

59. Newton Da Costa, Otávio Bueno, and Steven French, "Is There a Zande Logic?" *History and Philosophy of Logic* 19-1 (1998): 53.

60. Deleuze, *Différence et répétition*, 339; my transl.

61. Ibid., 245–47. The category is inspired by Leibniz. See, two decades after *Différence et répétition*, Deleuze, *Le Pli* (Paris: Minuit, 1988), 77–80.

62. Deleuze, *Différence et répétition*, 285; my transl.

63. Ibid., 36.

64. Deleuze, *Logique du sens* (Paris: Minuit, 1969), 49; my transl.

65. Ibid.

66. Ibid.; cf. 83.

67. Euripides, *Heracles*, ll. 1341–46; my transl. For line 1345, I restore the original (with *ontōs* instead of the common emendation *orthōs*).

68. Xenophanes in Diels and Kranze, eds., *Fragmente der Vorsokratiker*, B 11, my transl.; cf. B 12.

69. Ibid., B 14.

70. Karl Popper, *The World of Parmenides* (New York: Routledge, 1998), 95; Theodor Adorno and Max Horkheimer, *Dialectic of Enlightenment* (Stanford, Calif.: Stanford University Press, 2002), 2–4; Paul Feyerabend, *Farewell to Reason* (New York: Verso, 1987), 90–102.

71. Charles Baudelaire, *The Flowers of Evil,* in the 1954 transl. by William Aggeler (Fresno, Calif.: Academy Library Guild, 1954), 265 (with corrections for the last line). The original reads, "Je suis la plaie et le couteau! / Je suis le soufflet et la joue! / Je suis les membres et la roue, / Et la victime et le bourreau!"

72. Baudelaire, *Poems* (London: Harvill Press, 1952), transl. Roy Campbell, 107.

73. Baudelaire, *Les fleurs du mal* (New York: Washington Square Press, 1962), transl. George Dillon.

74. Transl. by Aggeler, 265; transl. of Baudelaire by Keith Waldrop, *The Flowers of Evil* (Middletown, Conn.: Wesleyan University Press, 2006), 105.

75. Transl. by Aggeler.

76. Transl. by Aggeler, with corrections.

77. *Ou* in French corresponds to the English "or" and is pronounced [u]. This alliteration, then, is at odds with the insistence on *and (et),* and it generates another logical disturbance.

78. I am building here on a suggestion by Aimable Dufatanye

(personal communication, July 2013). I am using the Sheffer stroke to mark mutually incompatible propositions.

79. $\alpha | \beta =_{def} \neg(\alpha \wedge \beta)$.

80. More generally, the last three stanzas of Baudelaire's "Heautontimoroumenos" would consist in a series of assertions of the type $(\pi \wedge \rho) \wedge (\pi|\rho)$. This could also be expressed as $(\pi \wedge \rho) \wedge \neg(\pi \wedge \rho)$.

81. Or $(q' \wedge \neg q') \wedge \neg(q' \wedge \neg q')$.

82. $\neg^*\alpha =_{def} \neg\alpha \wedge \neg(\alpha \wedge \neg\alpha)$.

83. $[(\pi \wedge \rho) \wedge \neg(\pi \wedge \rho)] \neq [(\pi \wedge \rho) \wedge \neg^*(\pi \wedge \rho)]$.

84. This last line was transcribed as $[(p' \wedge \neg p') \wedge \neg(p' \wedge \neg p')]$. This is equivalent to $(p' \wedge \neg^* p')$.

85. To reproduce Jean-Louis Dessalles's formalism in *La pertinence et ses origines cognitives,* if we have two evaluations v_i and v_j of the same state of affairs s, in this case, $v_i(s)\, v_j(s) < 0 \Leftrightarrow v_i(s) \uparrow v_j(s)$ [cognitive dissonance].

86. Augustine, *Confessions,* III.

87. Cf. Viktor Shklovski, "Kunst als Verfahren," in *Texte der russischen Formalisten,* Jurij Striedter, ed. (Munich: Fink, 1969–72), vol. 1, 2–35.

88. Classically, if an agent a believes α, we have the choice $(\neg\alpha \wedge aB\alpha) \vee (\alpha \wedge aB\alpha)$.

89. In modal logic, the *possible* and the *necessary* are respectively marked by $\Diamond$ and $\Box$.

90. Baruch Spinoza, *Tractatus de intellectus emendatione,* VIII, § 50; my transl.

91. $(a\Phi\alpha \wedge a\Phi\neg\alpha) \Leftrightarrow a\Phi(\alpha \wedge \neg\alpha)$.

II. ANIMAL MEDITATIONS

1. Plato, *Republic,* VI, 508c.

2. We could have, for instance, $R = \Delta r - nr + u$ *(realism)* and $R/n = r + u$ *(correlationism).*

3. Cf., in a different use, Ray Brassier, *Nihil Unbound: Enlightenment and Extinction* (London: Palgrave Macmillan, 2007), 239.

4. Luitzen Brouwer, *Collected Works* (New York: Elsevier, 1975), vol. 1, 490 and 524.

5. *Tao te ching,* poem 42.

6. Brouwer, *Collected Works,* vol. 1, 523.

7. Ibid.

8. Ibid., 421, for both quotes (my translation from the German).

9. Ibid., 523.

10. Ibid.

11. Ibid.

12. Alain Badiou, *Being and Event* (New York: Continuum, 2005), 8; cf. the original in *L'être et l'événement* (Paris: Seuil, 1988), 14.

13. Badiou, *Being and Event,* 40.

14. Badiou, *Logics of Worlds* (New York: Continuum, 2009), 53.

15. Ibid., 39.

16. Jacques Lacan, *Le Séminaire* (Paris: Seuil, 1973–), vol. 20, 107; my transl.

17. Ibid., vol. 17, 62.

18. Badiou, *Logic of Worlds,* 8.

19. Gilles Deleuze and Félix Guattari, *Qu'est-ce que la philosophie?* (Paris: Minuit, 1991), 144; my transl.

20. Ibid.

21. Henri Bergson, *L'évolution créatrice* (Paris: Alcan, 1908), 354; my transl.

22. Quentin Meillassoux, *Après la finitude* (Paris: Seuil, 2007), 111; my transl.

23. This is the main topic of René Descartes, *Meditationes,* III.

24. Xenophanes, in Diels-Kranz, *Fragmente,* B 15.

25. Rudolf Carnap, "The Elimination of Metaphysics through Logical Analysis of Language," in *Logical Positivism,* Alfred Ayer, ed. (Glencoe, Ill.: Free Press, 1959), 62.

26. Ibid., 63; my emphasis.

27. Ibid., 64.

28. Ibid., 66.

29. Ibid.

30. Ibid., 69. The equivalent expression Carnap uses throughout the original article is, in German, *logischer Mangel* ("Überwindung der Metaphysik durch logische analyse der Sprache," in *Erkenntnis,* 2, 1932, 229, etc.), which I'd rather translate as "logical lack" or "lack of logic." But because Carnap oversaw the English translation, and because "scientific philosophy" is unambiguous in its meaning, I am not going to comment any further.

31. See note to section 2, 80–81, of "Elimination of Metaphysics" in particular; the expression of *cognitive meaning* has been used by Carnap and others (but after the first publication of "Elimination") for a "referential" sense. Cf. also Moritz Schlick, "Positivism and Realism," in Ayer, *Logical Positivism,* 86–95.

32. Aristotle, *Metaphysics,* E 1, 1026a; my transl.

33. Ibid., A 2, 983a in particular.

34. Ibid., A 2, 982b.

35. Aristotle, *Metaphysics,* Γ 1, 1003a: *kath'auto* is a very difficult expression that has been diversely translated. In Plato, it functions as roughly equivalent to "as such" or "autonomously."

36. Cf. ibid., Γ 3, 1005a.

37. Alfred Jarry, *Gestes et opinions du Dr. Faustroll, pataphysicien* (end); my transl.

38. Anaxagoras in Diels-Kranz, *Fragmente,* B 12; my transl.

39. Steven Pinker, *The Blank Slate: The Modern Denial of Human Nature* (New York: Viking, 2002), 423.

40. Ibid., 424.

41. Ibid.

42. Ibid.

43. Ibid.

44. In *The Human Use of Human Beings: Cybernetics and Society* (Boston: Houghton Mifflin, 1954), 1, Norbert Wiener wrote, "To those of us who are engaged in constructive research and in invention, there is a serious moral risk of aggrandizing what we have accomplished."

45. David Chalmers, *The Conscious Mind: In Search of a Fundamental Theory* (New York: Oxford University Press, 1996), 293. The conjecture is presented as a sort of half-provocation by the author (see 310 especially).

46. Ray Kurzweill, *The Singularity Is Near: When Humans Transcend Biology.*

47. Emily Dickinson, *The Poems* (Cambridge, Mass.: Belknap Press, 1998), vol. 2, 595.

48. Protagoras, in Diels and Kranz, eds., *Die Fragmente der Vorsokratiker* (1952), fragment B 1; my transl.

49. Cf. Charles Darwin, in *Notebooks, 1836–1844* (Ithaca, N.Y.: Cornell University Press, 1987), 539–40 (manuscript of *Notebook M,* 84e).

50. Diogenes Laertius, *Lives and Opinions of Eminent Philosophers,* II, 40.

51. Sophocles, *Antigone,* l. 332. I am commenting the whole chorus appearing on ll. 332–75.

52. Ibid., l. 365.

53. Ibid.

54. Pat Shipman, *The Animal Connection: A New Perspective on What Makes Us Human* (New York: W. W. Norton, 2011).

55. Aristotle, *Politics,* A 2.

56. Isocrates, *Antidosis,* 253–55.

57. Cf. Emmanuel Levinas, *Totalité et infini* (The Hague: Nijhoff, 1961), chapter 7, 244–45.

58. Ekkehard König and Peter Siemund (with Stephan Töpper), "Intensifiers and Reflexive Pronouns," in *The World Atlas of Language Structures Online,* Matthew S. Dryer and Martin Haspelmath, eds. (Leipzig: Max Planck Institute for Evolutionary Anthropology, 2013), http://wals.info/chapter/47. See Shankara Bhat, *Pronouns* (New York: Oxford University Press, 2004).

59. Alan Turing, "Computing Machinery and Intelligence," *Mind* 59 (1950): 449. Interestingly, in 1956, Claude Shannon and John McCarthy already criticized this position, saying that an approach in terms of memory and response retrieval, that is, "Turing's definition of thinking," "does not reflect our usual intuitive concept of thinking"; see Shannon and McCarthy, eds., *Automata Studies* (Princeton, N.J.: Princeton University Press, 1956), vi.

60. Mario Tokoro and Luc Steels, eds., *A Learning Zone of One's Own: Sharing Representations and Flow in Collaborative Learning Environments* (Amsterdam: IOS Press, 2004), 144–47.

61. Donna Haraway, in Joel Weiss et al., eds., *International Handbook of Virtual Learning Environments* (New York: Springer, 2006), 117.

62. Cf. Ibid., 129, 147.

63. It could well be the case that what is known in computer science as the "*P* and *NP* problem" reflects the *cognitive* understanding of the difficulty I just outlined (if $P \neq NP$).

64. Immanuel Kant, *Logik,* "Einleitung, iii"; my transl.

65. Cf. Ashley Thompson, *Calling the Souls: A Cambodian Ritual Text* (Phnom Penh: Reyum, 2005).

66. Larry Squire et al., eds., *Fundamental Neuroscience* (Boston: Academic Press, 2008), 1081.

67. Plato, *Phaedo,* 85d.

68. Ibid., 86d.

69. Ibid., 92a–e.

70. Ibid., 66c.

71. Ibid., 70c–72e.

72. Ibid., 80b.

73. http://www.brainpreservation.org/. See Hans Moravec, *Mind Children: The Future of Robot and Human Intelligence* (Cambridge, Mass.: Harvard University Press, 1988).

74. http://eversmarterworld.wordpress.com/.

Repertory

This repertory does not give an exhaustive list of all the materials I consulted. Many references are selected for the overview they provide; other articles or books are much more idiosyncratic and, I believe, particularly valuable. I tend to put an emphasis on more recent and "noncanonical" works (noncanonical at least for someone trained in Continental philosophy and/or theory). Bibliographical records are only given for direct quotations, journal or Web articles, and book chapters in edited volumes. I also use the repertory to provide some additional content or more marginal reflections relating to the rest of the text.

1.

Disciplines. Peggy Kamuf, *The Division of Literature: Or the University in Deconstruction.* Laurent Dubreuil, "If Interdisciplinarity Means," *Sans Papier* (November 2007), http://as.cornell.edu/departments/french-t/publications/index7321.html?pubid=3886.

Historicity of the cognitive. Throughout the book, I mainly use the phrase "cognitive sciences" in the plural, because there is no unification at this point. In a few instances, I opt for the singular form of *science,* mainly for polemical reasons and to allude to a dogmatic approach. "Cognitivism" refers to the hard position dominating the first decades of cognitive studies (up to the early 1980s): the central role of artificial intelligence and quasi-religion of the machine, minimal concern with embodiment, theoretical priority given to *representation* and *computation.* Cognitivism certainly persists in today's science (often in a more fragmented form), though its phraseology is more and more diffuse now. As will become clearer in the rest of the book, I make a distinction between the "cognition" enacted by mental operations and "the cognitive," that is, a particular mode of thinking that promotes rationalistic enfolding, as the best or the most relevant one. In this sense, the "cognitive" (and the associated epistemic discourse) obviously predates by far the apparition of the cognitive sciences. See: Science Research Council, *Artificial Intelligence: Critical Concepts* (London: 1973), 35–37; Paul Thagard, "Welcome to the Cognitive Revolution," *Social Studies of Science* 19-4 (1989): 653–57; Pierre Steiner, "Cognitivisme et sciences cognitives," *Labyrinthe: Atelier Interdisciplinaire* 20 (2005): 13–39; Jean-Pierre Dupuy, *The Mechanization of the Mind*; Jamie Cohen-Sole, "Instituting the Science of Mind," *British Journal for the History of Science* 40-4 (2007): 567–97; David Golumbia, *The Cultural Logic of Computation.*

2.

Computer and the brain. Even von Neumann, writing just before his death in 1956, admitted some difficulties with the analogy; see the draft of his projected lectures at Yale, "aptly" titled *The Computer and the Brain* (New Haven, Conn.: Yale University Press, 1958), 81. In more technical publications of the exact same era, von Neumann was actually more straightforward, stating, for instance, "It is dangerous to identify the real physical (or biological) world with the models which are constructed to explain it. The problem of understanding the animal nervous action is far deeper than the problem of understanding the mechanism of a computer machine. Even plausible explanations of nervous reaction should be taken with a very large grain of salt." ("Probabilistic Logics," chapter 3 of Claude Shannon and John McCarthy, eds., *Automata Studies*, 96.) This warning should be renewed today; cf. also Bernard Stiegler, *La Technique et le temps* vol. 2, chapter III, §16. As for the current reconfiguration of thought processes in an interconnected and "digital" environment, see Nicholas Carr, *The Shallows: What the Internet Is Doing to Our Brains*; Eyal Ophir, Clifford Nass, and Anthony D. Wagner, "Cognitive Control in Media Multitaskers," *Proceedings of the National Academy of Sciences of the United States of America* 106 (2009): 15583–87; and Frédéric Kaplan, "How Books Will Become Machines," in *Lire demain*, Claire Clivaz et al., eds. (Lausanne: Presses polytechniques et universitaires romandes), 27–44.

3.

Variability. Marcus Meinzer et al., "Neural Signatures of Semantic and Phonemic Fluency in Young and Old Adults," *Journal of Cognitive Neuroscience* 21-10 (2009): 2007–18. Ferath Kherif et al., "The Main Sources of Intersubject Variability in Neuronal Activation for Reading Aloud," *Journal of Cognitive Neuroscience* 21-4 (2009): 654–68; Klaus Hoenig et al., "Conceptual Flexibility in the Human Brain," *Journal of Cognitive Neuroscience* 20-10 (2008): 1799–814; G. Rajkowska and P. S. Goldman-Rakic, "Cytoarchitectonic Definition of Prefrontal Areas in the Normal Human Cortex (2)," *Cerebral Cortex* 5-4 (1995): 323–37.

Toward a critique of current research in neuroimagery. Tal Yarkoni et al., "Cognitive Neuroscience 2.0," *Trends in Cognitive Sciences* 14-11 (2010): 489–96; Stephen Nadeau, *The Neural Architecture of Grammar* (Cambridge, Mass.: MIT Press, 2012), 177; Edward Vul and Hal Paschler, "Voodoo and Circularity Errors," *Neuroimage* 62-2 (2012): 945–48; Craig Bennet and Michael Miller, "How Reliable Are the Results from Functional Magnetic Resonance Imaging?," *Annals of the New York Academy of Sciences* 1191 (2010): 133–55.

Shannon, Jakobson, etc. Bertrand Geoghegan, "From Information Theory to French Theory," *Critical Inquiry* 38-1 (2011): 96–126. Commentators are gradually rediscovering the fact that information theory and cybernetics were very much in view for a whole generation of French thinkers.

4.

Paradox of the now. Augustine, *Confessions,* XI, 14–28 (cf. Aristotle, *Physics,* IV, 6). Jacques Derrida, "Ousia et grammè" in *Marges* (Paris: Minuit, 1972).

Flux, discontinuity, approximation. Henri Bergson, *L'évolution créatrice*; and William James, *The Principles of Psychology.* Cf. Gaston Bachelard, *Essai sur la connaissance approchée* (Paris: Vrin, 1968), 28, saying that a right use of the discontinuous may be "our only chance to vectorize our knowledge" (my transl.). At one point, Bergson briefly mentions the role of "*interruption*" (*Évolution créatrice*, 220).

5.

Neural networks and reentry. This question, as well as the idea of "multilayer perceptrons" (with hidden units, plus some feedforward and feedback processes), is hotly debated; see Gary Marcus, *The Algebraic Mind: Integrating Connectionism and Cognitive Science,* for a critical discussion. While I am quite convinced by Edelman's descriptions, my main point, at the end of the paragraph, is about the loop induced by the performance of cognition. Neural reentry would support the hypothesis at an anatomical level, and as Gerald Edelman wrote, it is something *more* than feedback: *The Remembered Present,* 65. But, overall, the cybernetic insistence on feedback (still vivid in Alicia Juarrero's *Dynamics in Action: Intentional Behavior as a Complex System*) as well as Douglas Hofstadter's suggestions (from *Gödel, Escher, Bach: An Eternal Golden Braid* to *I am a Strange Loop*) or the emphasis on biological self-reference in Humberto Maturana and Francisco Varela (*Autopoiesis and Cognition: The Realization of the Living,* as well as F. Varela, *Principles of Biological Autonomy*), and even older views about subjective reflexivity, are all possible takes for the enfolding movement I am beginning to describe.

6.

Cognition and emotion. Henri Laborit, *L'inhibition de l'action*; *Decoding the Human Message.* Antonio Damasio, *Descartes' Error: Emotion, Reason, and the Human Brain*; George Lakoff and Mark Johnson, *Philosophy in the Flesh: The Embodied Mind and Its Challenge to Western Thought*; Lawrence Barsalou, "Grounded Cognition," *Annual Review of Psychology* 59-1 (2008): 617–45; Luiz Pessoa, "Cognition and Emotion," *Scholarpedia* 4-1 (2009), http://dx.doi.org/10.4249/scholarpedia.4567; Frans De Waal, "What Is an Animal Emotion?" *Annals of the New York Academy of Sciences*

1224 (2011): 191–206; Baruch Spinoza, *Ethics*; Fernando Martinez-García, Amparo Novejarque, and Enrique Lanuza, "Evolution of the Amygdala in Vertebrates," in *Evolutionary Neuroscience,* Jon H. Kaas, ed. (Boston: Elsevier, 2009), 313–92.

7.

Complexity and science. Ming Li and Paul Vitányi, *An Introduction to Kolmogorov Complexity and Its Applications.* Henri Poincaré, *Science et méthode.* Edmund Husserl, *Krisis,* §9, is right in insisting on the mechanization of science (corresponding to what I situate in the cognitive), but he seems to reduce most—if not all—of science after Galileo to this repetitive technique, which is not true. (Cf. also François-David Sebbah, *Qu'est-ce que la technoscience?*) In *Blind Spot: Lectures on Logic,* Girard speaks of the "bureaucratic operations" (16) of mathematics and computation.

9.

General approach to complexity and emergence. Benoit Mandelbrot, *The Fractal Geometry of Nature*; Edward N. Lorenz, *The Essence of Chaos*; Ilya Prigogine and Isabelle Stengers, *Order Out of Chaos*; Mikhail Lyubich, "Regular and Stochastic Dynamics in the Real Quadratic Family," *Proceedings of the National Academy of Sciences of the United States of America* 95-24 (1998): 14025–27; Alwyn Scott, ed., *Encyclopedia of Nonlinear Sciences*; D. J. Watts and S. H. Strogatz, "Collective Dynamics of 'Small-World' networks," *Nature* 6684 (1998): 440–42; Philip Ball, *The Self-Made Tapestry: Pattern Formation in Nature.*

   History and values of the concept of emergence. Both Blitz, *Emergent Evolution,* and Peter Corning, *Holistic Darwinism: Synergy, Cybernetics, and the Bioeconomics of Evolution* (chapter 5) provide such a history. George Lewes is usually credited with the pioneering use of the term *emergent*; see his *Problems of Life and Mind* (London: Trübner, 1875), vol. 2, 146, 412–15. The mystical and religious component is strong in Pierre Teilhard de Chardin's rather early use of *emergence* and *self-organization,* or *auto-organisation* in French; see *The Phenomenon of Man* (New York: Harper, 1959), 220, 268–71, 305; or *Le phénomène humain* (Paris: Seuil, 1955), 244, 298–301, 338.

Scientific "complexity" and "chaos" in the humanities. Michel Serres, *Le passage du Nord-Ouest* (Paris: Minuit, 1980), 51, 61. Ilya Prigogine and Isabelle Stengers, *Order out of Chaos*; N. Katherine Hayles, *Chaos Bound*; K. Hayles ed., *Chaos and Order*; Gilles Deleuze and Félix Guattari, *Qu'est-ce que la philosophie?*; Robert Pepperell, *The Posthuman Condition: Consciousness beyond the Brain.* Achim Stephan, "The Dual Role of 'Emergence' in the Philosophy of Mind and in Cognitive Science," *Synthese* 151-3 (2006): 485–98. Gerald Vision, *Re-Emergence: Locating Conscious Properties in a Material World.*

Brain activity as a complex dynamic system. E. C. Zeeman, "Catastrophe Theory in Brain Modelling," *International Journal of Neuroscience* 6-1 (1973): 39–41. Gustavo Deco, Edmund T. Rolls, and Ranulfo Romo, "Stochastic Dynamics as a Principle of Brain Function," *Progress in Neurobiology* 88-1 (2009): 1–16. Dante Chialvo, "Emergent Complex Neural Dynamics," *Nature Physics* 6-10 (2010): 744–50. Scott Kelso, "Multistability and Metastability," *Philosophical Transactions of the Royal Society of London, Series B, Biological Sciences* 367-1591 (2012): 906–18. Harald Atmanspacher and Stefan Rotter, "Interpreting Neurodynamics," *Cognitive Neurodynamics* 2-4 (2008): 297–318; and, even more speculatively, Mikail Rabinovich et al., "Information Flow Dynamics in the Brain," *Physics of Life Reviews* 9-1 (2012): 51–73. Another nonclassical approach is based on quantum mechanics; see Roger Penrose, *The Emperor's New Mind: Concerning Computers, Minds, and the Laws of Physics*; Jack Tuszynski, ed., *The Emerging Physics of Consciousness*; Henry P. Stapp, *Mindful Universe: Quantum Mechanics and the Participating Observer.*

Bayesian approach. See also Thomas L. Griffiths, Charles Kemp, and Joshua B. Tenenbaum, "Bayesian Models of Cognition," in *The Cambridge Handbook of Computational Psychology,* Ron Sun, ed. (New York: Cambridge University Press, 2008), 59–100.

I firmly expect the scientific descriptions I sketch in this paragraph and throughout the whole book to become old-fashioned, or maybe irrelevant in the future. Thinking "with" the sciences includes the possibility of partial revisions, owing to the current limitations of experimental "results," mathematical developments, and theoretical underpinning. But I also believe the general reasoning of this book to be more durable, for most of its argument depends on a dialogue with the scientific rather than on an expansion of punctual content. This separates me from the endeavor of philosophers such as Evan Thompson. His *Mind in Life: Biology, Phenomenology, and the Sciences of Mind* (Cambridge, Mass.: Belknap Press, 2007) does a very remarkable job of presenting with a high level of detail and accuracy the contemporary theory of dynamic complex systems, and particularly in its relationship to mental phenomena. Such is not my goal at all. I foresee the revisable validity of emergentism, but neither do I feel the urgency to explain its tenets given the number of excellent books and articles already available on the topic—nor do I absolutely need these particular conjectures to be "true" to develop the hypothesis of a *beyond* that is provoked by *performance* (whatever its physical basis could be), especially through verbal language. Conceptually speaking, Thompson adopts a rather traditional stance in regard to his own "enactive" hypothesis: he is not in favor of abrupt reduction, but he would like the "gaps" to be bridged (*Mind in Life*, X, 6) and the "opacity" to be disposed of (ibid., 242). As I already said, and will repeat, the intellective even differs from the performance of cognition; it is precisely a way to confer signification to what is usually perceived as noisy interference within the realm of the epistemic (such as defects, gaps, or opacity).

11.

Matter and emergence. Maturana and Varela, *Autopoiesis and Cognition: The Realization of the Living*; Juarrero, *Dynamics in Action: Intentional Behavior as a Complex System*; Thompson, *Mind in Life: Biology, Phenomenology, and the Sciences of Mind*; Francis Bailly and Giuseppe Longo, *Mathematics and the Natural Sciences: The Physical Singularity of Life*; Terrence Deacon, *Incomplete Nature: How Mind Emerged from Matter*. There is a controversy on the overt or covert relationships between Deacon's essay and the books by Juarrero and Thompson. Deacon's personal contribution to a question that has already been largely discussed, from chemistry to philosophy, resides in an emphasis on "incompleteness." I certainly sympathize with such an interest in the negative. I do not think that, despite some lexical communality, Deacon and I construct very similar theories. Furthermore, the whole discussion about "teleodynamics" is quite foreign to me. A last remark: while emergence is tied to *matter,* it is not exactly reducible to its banal concept (and all the more in its initial conditions), so there is little hope to both maintain emergence and preach for *traditional* materialism. This is in resonance with Wiener's blunt observation: "Information is information, not matter or energy. No materialism which does not admit this can survive at the present day." *Cybernetics* (Cambridge, Mass.: MIT Press, 1948), 155. Despite their differences, the concepts of *emergence, autopoiesis,* and *information* (and maybe the discredited *Gestalt* as well) belong to scientific endeavors searching to renew (or go past) materialism. Other categories, tied to similar aims, will certainly appear later.

12.

Gödel's results. In the 1930s, both Jean Cavaillès, *Oeuvres complètes de philosophie des sciences* (Paris: Hermann, 1994), 144–59, and Albert Lautman, *Nouvelles recherches sur la structure dialectique des mathématiques* (Paris: Hermann, 1939), 17, interpreted the argument in relationship with the liar's paradox and the principle of noncontradiction. J. R. Lucas, in his "Minds, Machines, and Gödel," *Philosophy* 36-137 (1961), 112–27, used a similar argument to respond to both Turing's claims and the hopes of cybernetics. Three very interesting and contemporary formulations of the original theorems appear in Girard, *Blind Spot: Lectures on Logic,* chapter 2; in Bailly and Longo, *Mathematics,* chapter 2; and in Penrose, *Emperor's New Mind: Concerning Computers, Minds, and the Laws of Physics,* chapter 4, and *Shadows of the Mind: A Search for the Missing Science of Consciousness,* chapters 2–3. Penrose, as well as Bailly and Longo, provide impressive expansions of Gödel's argument within the sciences but outside of mathematics; see also Hofstadter, *Gödel, Escher, Bach:*

An Eternal Golden Braid, 696–719. In 1961, Jacques Derrida wrote at length about these theorems in his discussion of Husserl's work on mathematics, in Edmund Husserl, *L'origine de la géométrie* (Paris: Presses universitaires de France, 1962), 39–45, before explicitly referencing his own category of "undecidable" to Gödel a decade later in his *Dissemination.* Gilles Deleuze and Félix Guattari tie this same category to the "language of the mathematicians" in *L'Anti-Oedipe* (Paris: Minuit, 1972), 96. Graham Priest links his conception of *dialetheia* with Gödel's theorems in *In Contradiction: A Study of the Transconsistent* (New York: Oxford University Press, 2006), 46–48. See also Thom, *Modèles mathématiques,* 150–52.

13.

Nescience. Herbert Spencer, *First Principles,* § 4. Karl Popper, *Objective Knowledge,* 243–44. André Weil, "De la métaphysique aux mathématiques," in *Oeuvres scientifiques* (New York: Springer, 1979), vol. 2, 408–9. Manuel de Diéguez, *Science et nescience.* Laurent Dubreuil, "A Viral Lexicon for Future Crises," *Qui Parle* 20-1 (2011): 169–78. My own take has more to do with Bataille's *non-savoir*; see Georges Bataille, *Oeuvres complètes* (Paris: Gallimard, 1970–88), vol. 12, 278–88.

14.

Indiscipline. Laurent Dubreuil, ed., "La Fin des disciplines?," *Labyrinthe: Atelier interdisciplinaire* 27 (2007); "What Is Literature's Now?" *New Literary History* 38-1 (2007); and *L'état critique de la littérature.*

15.

Shared space. Mark Turner speaks of "a *virtual* brain distributed in the individual brains of all the participants in the culture" in *The Literary Mind: The Origins of Thought and Language* (New York:

Oxford University Press, 1996), 160. Gilles Châtelet, in both *Figuring Space: Philosophy, Mathematics, and Physics* and *L'enchantement du virtuel*, developed the category of the virtual, and his reflection played a role in Gilles Deleuze's later writings. After having completed the first draft of this book, I came to read Vittorio Gallese, "The Manifold Nature of Interpersonal Relations," *Philosophical Transactions of the Royal Society of London, Series B, Biological Sciences* 358-1431 (2003): 517–28, or "The 'Shared Manifold' Hypothesis," *Journal of Consciousness Studies* 8-5 (2001): 33–50. Gallese's very interesting ideas are in line with the current rediscovery of Husserlian intersubjectivity. I need to stress that, for me, the intellective space is created ad hoc (which implies it can disappear quasi-instantaneously) and that sharing is also *intra*-subjective.

16.

Dialogic. In classical Greek, *dialogikos* refers to dialogue and *dialogē* to an estimation, but as I said before, the prefix *dia* (as in *dianoia*) also indicates a process or an exchange. Priest, with *In Contradiction: A Study of the Transconsistent,* uses *dialetheia* (*dia* + *alētheia,* "truth") for somehow true contradictions. Floyd Merrell, *Sign, Textuality, World* (Bloomington: Indiana University Press, 1992), 77, uses *dia-logic* for "a temporally self-referential logic of the *other* according to which 'P' is not simply transformed into either 'not-P' or 'Q.'" I prefer to avoid *dialectics,* both because of its restrictive use in Aristotle (dialectics only combines well-received opinions and neutralizes the "what for"; see *Topics,* I, 1, 100b) and because of the Hegelian mark.

Affirmation and beyond. Jacques Derrida, *Ulysse Gramophone: Deux mots pour Joyce; The Work of Mourning.* Graham Priest, *Beyond the Limits of Thought.*

17.

Altered consciousness. J. D. Lewis-Williams et al., "The Signs of All Times," *Current Anthropology* 29-2 (1988): 201–45. Yuval Nir and

Giulio Tononi, "Dreaming and the Brain," *Trends in Cognitive Sciences* 14-2 (2010): 88–100. The cognitive and psychiatric effects of sensory deprivation have been largely studied in the United States, especially in the 1950s and 1960s. In the arts: Laurent Dubreuil, *De l'attrait à la possession.*

Madness. Erasmus, *Praise of Folly.* Michel Foucault, *Madness and Civilization: A History of Insanity in the Age of Reason.* Deleuze and Guattari, *L'Anti-Oedipe.*

20.

Connectome and graphs. Arthur Toga et al., "Mapping the Human Connectome," *Neurosurgery* 71-1 (2012): 1–5. Edward Bullmore and Danielle S. Bassett, "Brain Graphs," *Annual Review of Clinical Psychology* 7-1 (2011): 113–40.

Society, cognition, communication. Hal Whitehead, *Sperm Whales: Social Evolution in the Ocean* and *Analyzing Animal Societies: Quantitative Methods for Vertebrate Social Analysis.* F. A. Champagne and J. P. Curley, "Maternal Care as a Modulating Influence on Infant Development," in *Oxford Handbook of Developmental Behavioral Neuroscience,* Mark Blumberg et al., eds. (New York: Oxford University Press, 2010), 323–41. Charles Snowdon and Martine Hausberger, eds., *Social Influences on Vocal Development.* Dorothy Cheney and Robert Seyfarth, *Baboon Metaphysics: The Evolution of a Social Mind,* chapter 12. Klaus Zuberbühler "Linguistic Prerequisites in the Primate Lineage," in *Language Origins: Perspectives on Evolution,* Maggie Tallerman, ed. (New York: Oxford University Press, 2005), 262–82. Morten Christiansen and Nick Chater, "Language as Shaped by the Brain," *Behavioral and Brain Sciences* 31-5 (2008): 489–509, make the case that social coordination (which they call "C-induction") is easier to master than modeling the "natural world" ("N-induction"). A social and political origin of human language is a thesis notably developed by Robin Dunbar, *Grooming, Gossip, and the Evolution of Language*; Jean-Louis Dessalles, *Why We Talk: The Evolutionary Origins of Language.* The influence of writing on human thought is explored in Walter Ong, *Orality and Literacy.*

Neurons. Suzana Herculano-Houzel, "The Human Brain in

Numbers," *Frontiers in Human Neuroscience* (2009), doi:10.3389/
neuro.09.031.2009, and "The Remarkable, Yet Not Extraordinary,
Human Brain," *Proceedings of the National Academy of Science of the
United States of America,* 109-suppl. 1 (2012): 10661-8; F. A. Azevedo
et al., "Equal Numbers of Neuronal and Nonneuronal Cells," *Journal of Comparative Neurology* 513-5 (2009): 532–41; James Rilling,
"Differences between Chimpanzees and Bonobos," *Social Cognitive and Affective Neuroscience* 7-4 (2012): 369–79.

Sharing. Tomasello makes a great deal of "shared intentionality"
in *Origins of Human Communication,* 326, and *Why We Cooperate;*
this should be articulated with Herbert Clark's emphasis on "joint
activity" in human discourse (cf. his *Using Language*). Both in the
wild and in the lab, regular bonobos (without language) seem to
be more inclined to share food than chimpanzees are; see Rilling
et al., "Chimpanzees and Bonobos," 7–8, but see Craig Stanford in
Michael Arbib, ed., *Action to Language via the Mirror Neuron System*
(New York: Cambridge University Press, 2006), 102–6. My visits
to the bonobos of the lab facility then called the Great Ape Trust
in Des Moines, Iowa, took place in September 2010 and November
2011.

"The Animal." Jacques Derrida, *The Animal That Therefore I Am;*
The Beast and the Sovereign, 2 vols.

22.

Symbolic humans. Randall White, "Rethinking the Middle/Upper
Paleolithic Transition," *Current Anthropology* 23-2 (1982): 85–108.
Terrence Deacon, *The Symbolic Species: The Co-evolution of Language and the Brain* (New York: W. W. Norton, 1997), 340–74 in
particular. William Noble and Iain Davidson, *Human Evolution,
Language, and Mind: A Psychological and Archaelogical Inquiry.*
Ian Tattersall, *Becoming Human: Evolution and Human Uniqueness*
(New York: Harcourt Brace, 1998), 186, 226–28. Ángel Rivera Ar-
rizabalaga, *Arqueología del lenguaje.* Christopher Stuart Henshil-
wood and Francesco D'Errico, eds., *Homo Symbolicus: The Dawn
of Language, Imagination, Spirituality.*

23.

Primate communication. Dorothy Cheney and Robert Seyfarth, *How Monkeys See the World: Inside the Mind of Another Species* (Chicago: University of Chicago Press, 1990), 158–60 and 310 in particular. Laidre, "Meaningful Gesture in Monkeys?" *PloS One* 6-2 (2011) argues that, in a group of captive Mandrill, a gesture that has been invented by a female (covering one's eyes with one or two hands) is semantically understood and produced by other individuals. More generally, there is a thick bibliography about the advent of human verbal language through a repertoire of gestures; see, for instance, Michael Corballis, *From Hand to Mouth: The Origins of Language*; Willard van Orman Quine, *Word and Object.*

24.

Apes and language. Laurent Dubreuil, "La grande scène des primates," *Labyrinthe: Atelier interdisciplinaire* 38 (2012): 81–102.

25.

Defining language and meaning. Marc Hauser, Noam Chomsky, and Tecumseh Fitch, "The Faculty of Language," *Science* 298-5598 (2002): 1569–79. Jerry Fodor, *The Language of Thought; In Critical Condition; LOT* 2. Herbert Clark, *Arenas of Language Use.* Jean-Louis Dessalles, *Why We Talk: The Evolutionary Origins of Language.*

"Good enough" and algorithmic comprehensions of language. Anthony Sanford and Patrick Sturt, "Depth of Processing in Language Comprehension," *Trends in Cognitive Sciences* 6-9 (2002): 382–86. Fernanda Ferreira, "The Misinterpretation of Noncanonical Sentences," *Cognitive Psychology* 47-2 (2003): 164–203; F. Ferreira and Karl G. D. Bailey, "Disfluencies and Human Language Comprehension," *Trends in Cognitive Sciences* 8-5 (2004): 231–37.

Zones of convergence. H. Damasio et al., "A Neural Basis for Lexical Retrieval," *Nature* 380-6574 (1996): 499–505. Kyle Simmons and Lawrence Barsalou, "The Similarity-in-Topography Principle," *Cognitive Neuropsychology* 20-3 (2003): 451–86. Michael Corballis, "Mirror Neurons and the Evolution of Language," *Brain and Language* 112-1 (2010): 25–35. Marta Ghio and Marco Tettamanti, "Semantic Domain-Specific Functional Integration for Action-Related vs. Abstract Concepts," *Brain and Language* 112-3 (2010): 223–32. Timothy Rogers and James McClelland, eds., *Semantic Cognition: A Parallel Distributed Processing Approach*. Nadeau, *The Neural Architecture of Grammar*.

26.

Double bind. Gregory Bateson, "Minimal Requirements," in *Steps to an Ecology of Mind: Collected Essays in Anthropology, Psychiatry, Evolution, and Epistemology*; Jacques Derrida, *Glas*; Francisco Varela and Joseph Goguen, "The Arithmetic of Closure," *Journal of Cybernetics* 8-3 (1978): 291–324.

28.

Dynamics. One could postulate that words relatively stabilize thought (through a recurrent use of units) and that this stability is undone by sentences and the need to produce semantically flexible utterances. The cognitive ability to focus is required to benefit from lingual stabilization: children and teenagers whose attention is impaired have all kinds of difficulties with language. M. Bellani et al., "Language Disturbances in ADHD," *Epidemiology and Psychiatric Sciences* 20-4 (2011): 311–15. Groups of words function as dynamical systems in their own right; see, among others, J. L. Elman, "On the Meaning of Words and Dinosaur Bones," *Cognitive Science* 33-4 (2009): 547–82. It has been suggested, by various authors, that dynamical semantics could be higher in poetry; cf. Jan Mukařovský, *On Poetic Language*.

30.

"Kluge." Gary Marcus, *Kluge: The Haphazard Evolution of the Human Mind*; cf., for a more technical approach, Anna Kinsella and Gary Marcus, "Evolution, Perfection, and Theories of Language," *Biolinguistics* 3-2 (2009).

31.

Social usage of human language. Émile Benveniste, *Problèmes de linguistique générale* (Paris: Gallimard, 1966–74), vol. 2, 95. Laurent Dubreuil, *Empire of Language: Toward a Critique of (Post)colonial Expression.* Paul Thibodeau and Lera Boroditsky, "Metaphors We Think With," *PloS One* 6-2 (2011). Mark Landau, Daniel Sullivan, and Jeff Greenberg, "Evidence that Self-Relevant Motives and Metaphoric Framing Interact to Influence Political and Social Attitudes," *Psychological Science* 20-11 (2009): 1421–27. George Lakoff and Mark Johnson, *Metaphors We Live By.*

Self-organization and robotic simulation. Luc Steels, *The Talking Heads Experiment* (Antwerp, Belgium: Laboratorium, 1999); "Experiments on the Emergence of Human Communication," *Trends in Cognitive Sciences* 10-8 (2006): 347–49. Pierre-Yves Oudeyer, *Self-Organization in the Evolution of Speech.* Simon Kirby, "The Evolution of Meaning-Space Structure through Iterated Learning," as well as Andrew Smith, "Language Change and the Inference of Meaning," in *Emergence of Communication and Language,* Caroline Lyon, Chrystopher L. Nehaniv, and Angelo Cangelosi, eds. (London: Springer, 2007).

Co-evolution of language, culture, and brain. Deacon, *The Symbolic Species: The Co-evolution of Language and the Brain*; Christiansen and Chater, "Language as Shaped by the Brain"; Nick Chater et al., "Restrictions on Biological Adaptation in Language Evolution," *Proceedings of the National Academy of Sciences of the United States of America* 106-4 (2009): 1015–20. Luc Steels, "Modeling the Cultural

Evolution of Language," *Physics of Life Reviews* 8-4 (2011): 339–56. Kenny Smith and Simon Kirby, "Cultural Evolution," *Philosophical Transactions: Biological Sciences* 363-1509 (2008): 3591–603. More generally, see Kevin Laland, John Odling-Smee, and Sean Miles, "How Culture Shaped the Human Genome," *Nature Review Genetics* 11 (2010): 137–48, and Peter Richerson and Robert Boyd, *Not by Genes Alone: How Culture Transformed Human Evolution.*

Granularity of language. Tallerman, ed., *Language Origins: Perspectives on Evolution.* In 1851, the mathematician Antoine Augustin Cournot already mentioned the granularity (or discontinuity) of language as both an organizational requirement of the system and an "essential defectuosity." *Oeuvres complètes* (Paris: Vrin, 1973–89), vol. 2, 210; my transl.

Semantics and development. Blumberg et al., *Oxford Handbook of Developmental Behavioral Neuroscience.* Alison Gopnik and Andrew Meltzoff, *Words, Thoughts, and Theories* (Cambridge, Mass.: MIT Press, 1997), 195 in particular. Michael Tomasello, *Constructing a Language: A Usage-Based Theory of Language Acquisition.*

From semantic vagueness to notions. In a large bibliography about vagueness, one could consult Timothy Williamson, *Vagueness;* Roy Sorensen, *Vagueness and Contradiction;* Nicholas Smith, *Vagueness and Degrees of Truth;* Dominic Hyde, *Vagueness, Logic, and Ontology;* Paul Égré and Nathan Klinedinst, eds., *Vagueness and Language Use;* and, of course, Quine, *Word and Object.* As for "notions," see Antoine Culioli, *Cognition and Representation in Linguistic Theory;* Laurent Dubreuil, *L'état critique de la littérature* (Paris: Hermann, 2009), 34–43. Cf. René Thom, *Morphogenèse et imaginaire,* 81–84. Liane Gabora and Diederik Aerts, "Contextualizing Concepts Using a Mathematical Generalization of the Quantum Formalism," 2013, http://arxiv.org.proxy.library.cornell.edu/abs/1310.7682, develop an ingenious way of "computing" some aspects of meaning, but I would take exception with the widespread, though uncertain, terminology they use (converging ideation is not what I call a "concept").

Pirahã. Daniel Everett, "Cultural Constraints on grammar and cognition in Pirahã," *Anthropology* 46-4 (2005), 622, 632; "Pirahã Culture and Grammar," *Language* 85-2 (2009): 431. However, John Colapinto, in his reportage about the Pirahãs, notes that the

Pirahãs appear to enjoy and understand the movie *King Kong.* "The Interpreter," *The New Yorker,* April 16, 2007.

Experimental pragmatics and semiotics. Herbert Clark, *Semantics and Comprehension; Arenas of Language Use; Using Language.* Bruno Galantucci, ed., *Experimental Semiotics: Studies on the Emergence and Evolution of Human Communication.*

32.

Categories in verbal and nonverbal thought. Benveniste, *Problèmes,* vol. 1, chapter 6. Nicholas Evans and Stephen Levinson, "The Myth of Language Universals," *Behavioral and Brain Sciences* 32-5 (2009): 429–48. Zohar Eviatar and Raphiq Ibrahim, "Morphological Structure and Hemispheric Functioning," *Neuropsychology* 21-4 (2007): 470–84. Herbert Clark's *Arenas of Language Use* explores the dynamic of meaning construction in discursive interaction. Sue Savage-Rumbaugh (personal communication, 2011) posits an influence of differential language structures on the forms of human civilization. About friendhip, I draw here on some results coming from my *À force d'amitié.*

33.

Neuroimagery and language processing. N. Mashal et al., "An fMRI Investigation of the Neural Correlates Underlying the Processing of Novel Metaphoric Expressions," *Brain and Language* 100-2 (2007): 115–26. Zohar Eviatar and Marcel Just, "Brain Correlates of Discourse Processing," *Neuropsychologia* 44-12 (2006): 2348–59.

36.

Prosody. Lynne Nygaard, Debora S. Herold, and Laura L. Namy, "The Semantics of Prosody," *Cognitive Science* 33-1 (2009): 127–46. Reuven Tsur, *Toward a Theory of Cognitive Poetics.*

Literary "nevertheless." Dubreuil, "What Is Literature's Now?"

and *État critique*. For a different link between failure and the literary in a cognitive perspective, see Ellen Spolsky, *Gaps in Nature: Literary Interpretation and the Modular Mind* (Albany: State University of New York Press, 1993), as well as her article "Darwin and Derrida," *Poetics Today* 23-1 (2002): 43–62. Spolsky wants creativity to *bridge* the gaps (*Gaps,* 2), whereas the exhibition of the defective is for me what constitutes the opening toward the poetic intellective—if this was not clear enough, let me note here that my "intellective space" also refers to Maurice Blanchot's own *Literary Space* (or *L'espace littéraire,* whose infelicitous English title is, in its official translation, *The Space of Literature*).

Turns of minds. In more contemporary terms, one could speak of a combination of "intellectual styles" (sometimes called "thinking or cognitive styles," two expressions I would obviously tend to avoid) and "encoding strategies." See Engel Tulving and Donald Thomson, "Encoding Specificity and Retrieval Processes in Episodic Memory," *Psychological Review* 80-5 (1973): 352–73; Li-Fang Zhang and Robert Sternberg, eds., *The Nature of Intellectual Styles*; Michael Miller et al., "Individual Differences in Cognitive Style and Strategy Predict Similarities in the Patterns of Brain Activity between Individuals," *Neuroimage* 59-1 (2012): 83–93.

38.

Animal logic. Susan Hurley and Matthew Nudds, eds., *Rational Animals?,* chapter 10 by Josep Call ("Descartes' Two Errors") in particular; Alex Taylor et al., "Spontaneous Metatool Use by New-Caledonian Crows," *Current Biology* 17 (2007): 1504–7; Martin Schmelz, Josep Call, and Michael Tomasello, "Chimpanzees Know That Others Make Inferences," *Proceedings of the National Academy of Sciences of the United States of America* 108-7 (2011): 3077–9.

Logic in Asia. By way of introduction, see P. T. Raju, "The Principle of Four-Cornered Negation in Indian Philosophy," *Review of Metaphysics* 7-4 (1954): 694–713; Xy Jiang, "The Law of Non-contradiction and Chinese Philosophy," *History and Philosophy of Logic* 13-1 (1992): 1–14; François Jullien, *Detour and Access: Strategies of Meaning in China and Greece* or *The Silent Transformations*;

Richard Nisbett, *The Geography of Thought: How Asians and Westerners Think Differently ... and Why*.

In the link I am making between the geometric presentation of the macroscopic environment, I am influenced by the remarks of Thom (*Stabilité structurelle*, §13.1.A, 294; *Morphogenèse et imaginaire*, 40–45), even though I change the order of consequences he describes. See also Poincaré, *La science et l'hypothèse* (Paris: Flammarion, 1917), 109. I assume the reflexive empiricism marking my description of logicality; it could well be a sign of my own relation to "non-philosophy." See Derrida, Introduction to Husserl, *L'origine de la géométrie*, 109; *L'écriture et la différence*, 226. This position implies that, depending on the segment of reality to which we aim to relate, logical lines do not have the same accuracy or inefficiency. But logics are not secreted by reality; they are a matter of our mental negotiations with it (as we are locally making it real). Then, the different lines do not have to be contained in or restricted to this or that region of the real. This is how, for instance, formalized logics originally developed for quanta could be rightly used to describe a cognitive state even if it does not principally occur at the quantum level. The necessary link between a given formal approach and a physical state of matter is a problem when the material apparatus of thought is being examined. This drives Roger Penrose and Stuart Hameroff to postulate that microtubules act as quantum processing units. This last claim has been met with skepticism. But whatever its relevance could be, if the particular formalization of one specific line is, as I argue, a mental invention under the condition of perceived reality, then we can see the more or less acceptable "migration" of a logic or formalism from one setting to the other, without having to ratify physicalistic conjectures. See Liane Gabora and Diederik Aerts, "A Model of the Emergence and Evolution of Integrated Worldviews," *Journal of Mathematical Psychology* 53-5 (2009): 434–51; Newton Da Costa and Steven French, *Science and Partial Truth*; Roger Penrose, *Shadows of the Mind: A Search for the Missing Science of Consciousness,* chapters 5 and 7; Stuart Hameroff and Roger Penrose, "Orchestrated Reduction of Quantum Coherence in Brain Microtubules," *Mathematics and Computers in Simulation* 40-3 (1996): 453–80; Jeffrey Reimers et al., "The Revised Penrose–Hameroff Orchestrated Objective-Reduction

Proposal for Human Consciousness Is Not Scientifically Justified,"
Physics of Life Reviews 11-1 (2014): 101–3; Karen Barad, *Meeting the
Universe Halfway: Quantum Physics and the Entanglement of Matter
and Meaning*; Henry Stapp, *Mindful Universe: Quantum Mechanics
and the Participating Observer*; Gilles Châtelet, *L'enchantement du
virtuel* (Paris: Rue d'Ulm, 2010), 245–52.

39–40.

Contradiction and logic. Newton Da Costa, *Logique classique et non-
classique*; Jean Cavaillès, *Oeuvres complètes*.

41.

Inconsistency of language. Gottlob Frege, *Begriffsschrift*. The idea
of the inconsistency of language has become a minority position
among analytic philosophers. Matti Eklund is one of the last ortho-
dox in this matter; see "Inconsistent Languages," *Philosophy and
Phenomenological Research* 64-2 (2002): 251–75, or "Deep Inconsistency," *Australasian Journal of Philosophy* 80-3 (2002): 321–31. He
assigns a purely verbal origin to paradoxes. He further postulates
that linguistic negation is too weak or superficial to produce actual
contradictions, that, in any case, "speakers of English . . . abhor"
("Deep Inconsistency," 328). At most, then, paradoxes "exert pull"
("Inconsistent Languages," 252), whatever that could mean. An
obvious question is, Are we so sure that all these difficulties are
cancelled in formalized logic? If they are maintained to some
extent—as Gödel's theorems already assured, and as other mathematicians have argued (Brouwer, *Collected Works*, vol. 1, 421; Thom,
Modèles mathématiques, 150–51)—then we could make similar
claims vis-à-vis formal equations: that they are neither absolutely
consistent nor complete, that their negation is not total, and so on.
The defect of thinking is amplified through its performance, that
is, through itself; no correction is complete enough.

    A logic without identity. Plato, *Cratylus*, 439e–440e; Jacques
Derrida, *Limited Inc.* and *Paper Machine*, chapter 8; cf. Christopher

Norris, *Language, Logic, and Epistemology: A Modal-Realist Approach*, chapter 1, and *Fiction, Philosophy, and Literary Theory: Will the Real Saul Kripke Please Stand Up?*, chapter 1; David A. White, *Derrida on Formal Logic: An Interpretive Essay*, chapters 8–10; *The I Ching, or, Book of Changes*. There have been some attempts to formalize logic without an identity of identity: in addition to quantum logic postulating entanglement, we may quote the even more idiosyncratic (and relatively impenetrable) efforts of Décio Krause and Jean-Yves Béziau, "Relativizations of the Principle of Identity," *Logic Journal of the Interest Group in Pure and Applied Logics* 5 (1997): 327–38. See also Steven French and Décio Krause, *Identity in Physics: A Historical, Philosophical, and Formal Analysis*, or Graham Priest, "Non-transitive Identity," in *Cuts and Clouds: Vagueness, Its Nature, and Its Logic*, Richard Dietz and Sebastiano Moruzzi, eds. Girard's noncommutative level of formalized logic is also displacing identity (*Blind Spot: Lectures on Logic*, 410–14). In the dynamic model Lupasco proposed in the early 1950s—transcribing a kind of "quantum dialectic" between a potential event and an actual event—identity is also more fugacious (though not absent); see Stéphane Lupasco, *Le principe d'antagonisme et la logique de l'énergie*, and a more recent appropriation in Joseph Brenner, *Logic in Reality*.

43.

Logic ≠ Logic. Philip Johnson-Laird, *Mental Models: Towards a Cognitive Science of Language, Inference, and Consciousness.* Antoine Culioli, *Cognition and Representation in Linguistic Theory* (Philadelphia: Benjamins, 1995), 109–11. Jean-Louis Dessalles, *La pertinence et ses origines cognitives* (Paris: Hermès-sciences, 2008), chapters 5 and 6 in particular. Jesse Prinz, *Beyond Human Nature: How Culture and Experience Shape the Human Mind*, chapter 8. See also William Cooper, *The Evolution of Reason: Logic as a Branch of Biology.*

Nonstandard logic. Jan Łukasiewicz, "3-Valued Logic," "Determinism," "Many-Valued Systems," in *Polish Logic 1920–1939*, Storrs McCall, ed. (Oxford: Clarendon Press, 1967). Alfred Tarski, "Concept of Truth," "Concept of Logical Consequence," in *Logic*,

Semantics, Metamathematics: Papers from 1923–1938. Willard van Orman Quine, "What Price Bivalence," *Journal of Philosophy* 78-2 (1981): 90–95; Newton Da Costa, *Logique*; Graham Priest, *In Contradiction: A Study of the Transconsistent* and *An Introduction to Non-classical Logic: From If to Is.*

45.

Paraconsistence and dialetheism. Da Costa and French, *Science and Partial Truth: A Unitary Appraoch to Models and Scientific Reasoning* (New York: Oxford University Presss, 2003), 87; Berto, "Meaning, Metaphysics, and Contradiction."

Principle of noncontradiction in Aristotle and beyond. Jan Łukasiewicz, *Aristotle's Syllogistic from the Standpoint of Modern Formal Logic*; Stéphane Lupasco, *Logique et contradiction*; Graham Priest, *Doubt Truth to Be a Liar*; François Jullien, *Si parler va sans dire*; Patrick Grim, "What Is a Contradiction," in *The Law of Noncontradiction,* G. Priest, J. C. Beall, and B. Armour-Garb, eds. (Oxford: Clarendon Press, 2004): 49–72, lists dozens of different ways of formalizing the principle of noncontradiction.

46.

Also see Jean Bollack and Heinz Wismann, *Héraclite ou la séparation.* A profound (and somehow unexpected) comprehension of Heraclitus appears in the work of mathematician René Thom (throughout *Stabilité structurelle*; see in particular the remark in §1.4.D, 10; *Modèles mathématiques,* chapter 10).

47.

Contradictions in a colonial context. Lucien Lévy-Bruhl, *Fonctions mentales; L'âme primitive; Les carnets* (Paris: Presses universitaires de France, 1949), 159. Edward Evan Evans-Pritchard, "The Zande Corporation of Witchdoctors," *Journal of the Royal Anthropological*

Institute of Great Britain and Ireland 63 (1933): 63–100, and *Witchcraft, Oracles, and Magic among the Azande.* Laurent Dubreuil, *Empire of Language: Towards a Critique of (Post)colonial Expression,* 29–32. Newton Da Costa, Otávio Bueno, and Steven French, "Is There a Zande Logic?," 51.

Contradictions and literature. Maurice Blanchot, *Faux pas* and *La part du feu.* Jacques Rancière, *Mute Speech.* Laurent Dubreuil, *De l'attrait*; "Des raisons de la littérature," *Labyrinthe: Atelier interdisciplinaire* 14 (2003): 12–24; *L'état critique de la littérature.* Cf. also Roman Jakobson, "Poetry of Grammar," in *Verbal Art, Verbal Sign, Verbal Time*; Stéphane Lupasco, *Logique et contradiction* (Paris: Presses universitaires de France, 1947), 161–86 (but, here, literary contradictions are reduced to the affective and are severed from cognition).

53.

Nonconsistent sets. In 1920, Łukasiewicz suggested three values 0, 1, and 2 (in Storrs McCall, ed., *Polish Logic 1920–1939,* 18) before proposing, a decade later, to normalize the values—as is usually done now—with 0, 1/2, and 1 (*Polish Logic 1920–1939,* 54). Da Costa, "On a Theory of Inconsistent Formal Systems," *Notre Dame Journal of Formal Logic* 15-4 (1974): 497–510, explains the basis of C1. I am otherwise adapting Jean-Louis Dessalles, *La pertinence et ses origines cognitives,* chapter 6, with a few revisions, one being conceptually important: I stick to *incompatibility* for the negative value introduced by the product of different evaluations, instead of reintroducing "contradiction," as Dessalles does here and there (105 in particular, where "contradictory" is juxtaposed to "incompatibility"). On the formalization of logical incompatibilities, I used in particular Jacques Picard, "Les normes formelles du raisonnement déductif," *Revue de métaphysique et de morale* 45-2 (1938): 213–54; Robert Blanché, *Structures intellectuelles: Essai sur l'organisation systématique des concepts* (Paris: Vrin, 1966), 28; Amable Dufatanye, personal communication, July 2013. For the other formal systems used in this paragraph, see Graham Priest, *An Introduction to Nonclassical Logic: From If to Is.*

54.

Fiction and contradiction. See: Francesco Berto, "Modal Meinongianism and Fiction," *Philosophical Studies* 152-3 (2011): 313–34, for an overview of the issue in analytic philosophy; Gregory Currie, *The Nature of Fiction* (New York: Cambridge University Press, 1990), 177–80 against contradiction; Floyd Merrell, *Pararealities: The Nature of Our Fictions and How We Know Them* (Philadelphia: Benjamins, 1983), 24–25, who argues for an *"oscillatory model"* of fiction, for "it is logically impossible to perceive-imagine a foregrounded item from 'within' a fictional frame and *at the same precise instant* foreground an item *from* the 'real world'"; Ellen Spolsky, "Why and How to Take the Fruit and Leave the Chaff," *SubStance* 30-1 (2001): 177–98. Cf. Paul De Man, *Blindness and Insight: Essays in the Rhetoric of Contemporary Criticism* (New York: Oxford University Press, 1971), 17–19.

Neural basis of fiction. Demis Hassabis et al., "Patients with Hippocampal Amnesia Cannot Imagine New Experiences," *Proceedings of the National Academy of Sciences of the United States of America* 104-5 (2007): 1726–31. Demis Hassabis and Eleanor A. Maguire, "The Construction System of the Brain," *Philosophical Transactions of the Royal Society of London, Series B, Biological Sciences* 364-1521 (2009): 1263–67, and "Deconstructing Episodic Memory with Construction," *Trends in Cognitive Sciences* 11-7 (2007): 299–306.

55.

¬a ∧ aΦa. Graham Priest, *Towards Non-being: The Logic and Metaphysics of Intentionality,* uses, at its core, a formalization of this type.

Expert performance. Karl Anders Ericsson et al., ed., *The Cambridge Handbook of Expertise and Expert Performance* (New York: Cambridge University Press, 2006). Neil Charness et al., "The Role of Deliberate Practice in Chess Expertise," *Applied Cognitive Psychology* 19-2 (2005): 157.

See the role of confabulation and of the hypothetical "left-hemisphere interpreter" in split-brain research: M. Gazzaniga et al., "Collaboration between the Hemispheres of a Callosotomy

Patient," *Brain: A Journal of Neurology* 119-4 (1996): 1255–62. William Hirstein, *Brain Fiction*. Lionel Naccache is currently developing the category of fictionalization: "Visual Consciousness," in *The Neurology of Consciousness: Cognitive Neuroscience and Neuropathology*, Steven Laureys and Giulio Tononi, eds. (London: Academic, 2009), 277–78. Contemporary cognitive neuroscience may be rediscovering Hans Vaihinger's *Philosophy of "As If"* (without some of its radical conclusions). In his early reflection on experimental psychology, Maurice Merleau-Ponty also speaks of consciousness attending to the "spectacle entier du monde" in *La structure du comportement* (Paris: Presses universitaires de France, 1949), 204.

59–60.

Realism and beyond. Hilary Putnam, *Realism with a Human Face*; Drew Khlentzos, *Naturalistic Realism and the Antirealist Challenge*; James Ladyman, Don Ross et al., *Every Thing Must Go: Metaphysics Naturalized*; Barad, *Meeting the Universe Halfway: Quantum Physics and the Entanglement of Matter and Meaning*; Leni Bryant, Nick Srnicek, and Graham Harman, eds., *The Speculative Turn: Continental Materialism and Realism*; Isabelle Stengers, *Cosmopolitics 2*; Bruno Latour, *An Inquiry into Modes of Existence*; Tristan Garcia, *Form and Object*.

61.

Realism and mathematics. René Thom, *Stabilité structurelle*, §7.1.B, 122–23. Bailly and Longo, *Mathematics and the Natural Sciences: The Physical Singularity of Life*. Hans Primas was a direct inspiration for the category of "tenseless" truths. Primas, "Non-Boolean Descriptions for Mind-Matter Problems," *Mind and Matter* 5-1 (2007): 7–44. About numeration in animals, Damian Scarf, Harlene Hayne, and Michael Colombo, "Pigeons on Par with Primates in Numerical Competence," *Science* 334-6063 (2011): 1664. Stanislas Dehaene, *The Number Sense: How the Mind Creates Mathematics*.

65.

The gods are here to make sense of the no-sense. In a still unpublished manuscript, Meillassoux explicitly writes that "the absurdity of the world, once thought radically, ceases to be the locus of the most violent despair and becomes the very field of the most violent hope" (*L'inexistence divine*, doctoral diss., Université Paris-I, 1997, 6; my transl.) as soon as it is obvious that "the philosophers' God" (ibid., 372) will one day appear in the universe, in prelude to "*the real possibility of immortality*" (ibid., 290) that is our resurrection. Yippee. See Graham Harman, *Quentin Meillassoux: Philosophy in the Making* (Edinburgh: Edinburgh University Press, 2011), 110–21, and 221–38 for some excerpts and comments in English. Cf. Ray Brassier, *Nihil Unbound: Enlightenment and Extinction*, 85–96, on the "paradox of contingency."

73.

Greek materialism and idealism. See Aristotle, *Metaphysics*, A 3–6, for a description of, the debate, up to Plato.

75.

On the legacy of Protagoras. In a very different way, see how Isabelle Stengers appropriates Protagoras's motto and tries to both favor the figure of the "non-relative sophist": *Invention of Modern Science* (Minneapolis: University of Minnesota Press, 2000), 145; *Cosmopolitics* (Minneapolis: University of Minnesota Press, 2010), vol. 1, 11. She reassesses the issue of measurement through the experience of quantum physics in particular in her *Cosmopolitics*, vol. 2, 3–104.

78.

A note about technique. There are *plenty* of techniques in the world we live in, if we abide by a general view of "mechanical" life processes; even if we stick to a narrower conception, techniques (including the design and use of tools of whatever disposition) are to be found across animal *taxa*. I am certainly not "against technique"; this would be ridiculous. I am much, much more suspicious toward "techno-logy," that is, a *discourse* giving *reason* (*logos* in Greek, both times) to a necessary supremacy of technique. Why am I suspicious? Precisely because *technique* is thin. *Technology* is nothing more than a phraseology considering techniques as purely autonomous entities that we would theorize (like the Platonic soul does with Ideas) and realize for our own good. Nothing of the sort could be said: techniques, once examined *in vacuo*, are literally "valueless" (I did not say "neutral"); promoting or demoting them is consequently worthless. The recent confusion, among fashionable scholars, between *technique* and *technology*, giving way to categories like the "technology of the self" or the "technology of love," is simply regrettable (see how Michel Foucault's title *Techniques de soi* became in English *Technologies of the Self*, or the new cohort of essays similar to Dominic Pettman's *Love and Other Technologies: Retrofitting Eos for the Information Age*). When I use the Greek name *tekhnē* (as I did in paragraphs 31 and 34), I do so to introduce the possibility of a link between *tools, techniques,* and *artistic competence* (also close to the craft of an artisan) through *one* word. Please note that a link is not an equation; I am referring to gradients in terms of mental capacities and to large differences of intensity. I also use *poiēsis* for its semantic wealth, referring to production, fabrication, creation, and poetry. I am not opposing *poiēsis* to *tekhnē*: the possibility of the former comes from the existence of the latter, in a relationship that is either the emergence stemming from a dynamic system or the sudden introduction to the intellective through the performance of cognitive operations. If some readers still have doubts about the large incompatibility between the Heideggerian position and my own thesis, I invite them to check the comment of Sophocles's chorus in "The Ode on Man" (Martin Heidegger, *Introduction to Metaphysics*). Cf. also Bernard Stiegler, *La technique et le temps*, vol. 1, chapter II, §5, and 3, chapter VI, §6; Bruno Latour, *Enquête sur les modes d'existence* (Paris: La Découverte, 2012), 228–29, 248.

79.

Bipedalism. R. Alexander, "Bipedal Animals, and Their Differences from Humans," *Journal of Anatomy* 204-5 (2004): 321–30; Susana Carvalho et al., "Chimpanzee Carrying Behaviour and the Origins of Human Bipedality," *Current Biology* 22-6 (2012): 180–81.

Tools. Jane Goodall, *The Chimpanzees of Gombe: Patterns of Behavior,* chapter 18; B. Kenward et al., "Tool Manufacture by Naive Juvenile Crows," *Nature* 433-7022 (2005): 121; Itai Roffman et al., "Stone Tool Production and Utilization by Bonobo-Chimpanzees (*Pan paniscus*)," *Proceedings of the National Academy of Sciences of the United States of American* 109-36 (2012): 14500–3.

Language. Juliane Kaminski, Josep Call, and Julia Fisher, "Word Learning in a Domestic Dog," *Science* 304-5677 (2004): 1682–83. Roger Fouts and Stephen Tukel Mills, *Next of Kin.* Laurent Dubreuil, ed., *L'éloquence des singes,* special issue of *Labyrinthe: Atelier interdisciplinaire* 38 (2012).

Masks and adornment. Wolfgang Köhler, *The Mentality of Apes* (London: Routledge, 1973), 93–95, 316 (where the author even speaks of "the chimpanzee's passion for adorning himself"). Sue Savage-Rumbaugh, *Ape Language: From Conditioned Response to Symbol.*

Politics. Frans De Waal, *Chimpanzee Politics: Power and Sex among Apes.* Nigel Bennett and Chris Faulkes, *African Mole Rats: Ecology and Eusociality.*

Domestication. Brian Hare, Victoria Wobber, and Richard Wrangham, "The Self-Domestication Hypothesis," *Animal Behaviour* 83-3 (2012): 573–85. Cf. Peter Sloterdijk, *Nicht gerettet.*

80.

Ape language research. Sue Savage-Rumbaugh, *Ape Language: From Conditioned Response to Symbol*; Sue Savage-Rumbaugh, Stuart Shanker, and Talbot Taylor, *Apes, Language, and the Human Mind*; Pär Segerdahl, William Fields, and Sue Savage-Rumbaugh, *Kanzi's Primal Language: The Cultural Initiation of Primates into Language*;

Sue Savage-Rumbaugh, "Human Language—Human Consciousness," http://onthehuman.org/2011/01/human-language-human
-consciousness/. For a standard historical overview on ape language research, see Gregory Radick, *The Simian Tongue: The Long
Debate about Animal Language*; for a nonstandard one, see Dubreuil, ed., *L'éloquence des singes*.

82.

Levinas. Emmanuel Levinas, *Totalité et infini,* 52–53 (face-to-face),
175–90 (language and the Other), 254–57 (biology and society).

83.

Mirrored self. Charles Darwin, *The Expression of the Emotions in
Man and Animals* (New York: Appleton, 1886), 142. Jacques Lacan,
"The Mirror Stage," in *Écrits: A Selection* (New York: W. W. Norton,
1977), 1–7. Maurice Merleau-Ponty, *Merleau-Ponty à la Sorbonne*
(Grenoble: Cynara, 1988), 318ff.; Bruno Cassou-Noguès, *Le bord
de l'expérience* (Paris: Presses universitaires de France, 2010), 155.
Kalina Christoff et al., "Specifying the Self for Cognitive Neuroscience," *Trends in Cognitive Sciences* 15-3 (2011): 104–12. James An-
derson and Gordon Gallup, "Which Primates Recognize Themselves in Mirrors?," *PLoS Biology* 9-3 (2011). Tetsurō Matsuzawa,
ed., *Primate Origins of Human Cognition and Behavior* (New York:
Springer, 2001), 297–404. In a nonsystematic fashion, Köhler,
in the 1910s, already introduced mirrors to chimpanzees (and to
what I believe to be bonobos, such as the ape named Koko) and
noticed reactions that are similar to what more recent experiments
describe (*The Mentality of Apes,* 317–19). Dan Zahavi's remarkable
work on the self is infused with both contemporary psychology
and German and French phenomenology; see, for instance, his
Subjectivity and Selfhood: Investigating the First-Person Perspective.
Also see Uriah Kriegel, ed., *Self-Representational Approaches to
Consciousness.*

Alternative approach to mirror recognition. One could evoke here
the case of the right hand touching the left hand (after a famous,
and most discussed, idea of Edmund Husserl, *Ideas,* 2, § 36). There
are two distinct experiences here. One contributes to the formation of a body "image"—or mental map. We have no reason to believe that a cat is unable to grasp, at least at some minimal level, the
fact that it is *this* body, by repeatedly touching or licking itself. The
other experience of self-touch, as inducing reflexivity, is apparently
inaccessible without mirror recognition—or we would find more
animal species being capable of this feat. Among members of *Homo
sapiens,* social institutions and language provide the category of the
self, no matter what. As a result, the *individual* experience becomes
much less crucial. In other terms, through social usage, the blind
could see themselves through touch and discourse. However, I
need to add that some experiments with human subjects tend to
confirm the actual limit of cognitive reflexivity in the absence of
sight.

84.

A verbal self. Benveniste, *Problèmes,* vol. 1, chapters 20 and 21; Dan-
iel Dennett, *Brainchildren: Essays on Designing Minds,* chapter 24.
Gerard Edelman, *The Remembered Present: A Biological Theory of
Consciousness,* 96–103; Gerard Edelman and Giulio Tonioni, *A
Universe of Consciousness: How Matter Becomes Imagination,* chap-
ter 4; Paul Churchland, *The Engine of Reason, the Seat of the Soul:
A Philosophical Journey into the Brain* (Cambridge, Mass.: MIT
Press), 269–71.

86.

Nature and culture. Jesse Prinz, *Beyond Human Nature: How Culture and Experience Shape the Human Mind*; Serge Margel, *Logique
de la nature.*

87.

Relaxation of sexual selection. Terrence Deacon, "Relaxed Selection," in *Oxford Handbook of Developmental Behavioral Neuroscience,* Blumberg, ed., chapter 35.

Cumulative culture. André Leroi-Gourhan, *Gesture and Speech* (Leroi-Gourhan's reflection inspired the category of the epiphylogenetic in the first volume of Bernard Stiegler, *Technique et le temps*); Andrew Whiten, "The Scope of Culture in Chimpanzees, Humans, and Ancestral Apes," *Philosophical Transactions of the Royal Society of London, Series B, Biological Sciences* 366-1567 (2011): 997–1007.

Machines and robots. Julien Offray de La Mettrie, *Man a Machine.* Gilbert Simondon, *Du mode d'existence des objets techniques; L'individuation à la lumière des notions de forme et d'information.* Luc Steels, "Intelligence with Representation," *Philosophical Transactions of the Royal Society of London, Series A, Mathematical, Physical, and Engineering Sciences* 361-1811 (2003): 2381–95.

Creativity. There is some possible evidence for the role of the right hemisphere of the brain in "creativity" (for left-lateralized subjects). See Richard Chi and Allan Snyder, "Facilitate Insight by Non-invasive Brain Stimulation," *PloS One* 6-2 (2011); B. Miller et al., "Emergence of Artistic Talent in Frontotemporal Dementia," *Neurology* 51-4 (1998): 978–82; J. Kounios and M. Beeman, "The Aha! Moment," *Current Directions in Psychological Science* 18-4 (2009): 210–16; but also Gwenda Schmidt et al., "Beyond Laterality," *Journal of the International Neuropsychological Society* 16-1 (2010): 1–5. This should not be viewed as going against the "strong position" I am advocating. As a matter of fact, it *may* appear that, roughly speaking, the right hemisphere of the brain is solicited more than usual when unexpected information or situations are being processed: it would serve the role of an additional force. As we said before (§33), according to one of the rare studies on the topic, "poetic metaphors" would require a partial activation of both lobes, in contradistinction with random association of words and widespread metaphors. Unless we want to attribute unique properties to the right hemisphere (and, given neural plasticity as we currently know it, this would not be wise), these few experiments would lead me to believe that abandoning a mental routine or having an insight depends on a capacity to augment parallel distribution in the brain. Precisely, I would advance that the intellective journey displaces us and asks from us a new and intense intellectual effort—as for trials and errors, they are just *one* way to address the appearance of novel ideas, contrary to what Jean Pierre Changeux repeats; see, for instance, his "Creation, Art, and the Brain," in *Neurobiology of Human Values,* J. P. Changeux et al., ed. (New York: Springer, 2005), 1–10. For a better and "cognitive" view about the content of my strong position, read Graeme Halford, William Wilson, and Steven Philips, "Relational Knowledge," *Trends in Cognitive Sciences* 14-11 (2010): 497–505, presenting a conjectural integration of heuristic and analytic thinking. Nonetheless, this new model, for me, still occults the outside of cognition. The "strong Artificial Intelligence" approach to creativity (conflating invention with *poiēsis,* keeping the mental within, and oversimplifying everything) has been sterile so far. See the gap between the achievements and the claims of R. Schanck or others in Steven Smith, Thomas Ward, and Ronald Finke, eds., *The Creative Cognition Approach,* as well as the very sensible remarks by Roger Penrose, *Shadows of the Mind: A Search for the Missing Science of Consciousness* (New York: Oxford University Press, 1994), 154–56, 393, 399–401, and further precisions on this topic by Gilles Châtelet, *L'enchantement du virtuel,* 215–32, 245–52.

Two final remarks on "Emily Howell." One, Cope does a lot of "human" work of selection and articulation when he enters information in his software. Basically, a ready-made understanding of music is *given* to the computer, which will mainly act as a calculator, with lots of pitfalls. In this respect, we have more to deal with computer-assisted creation than anything else. Then, the work of François Pachet at the Sony Lab in Paris on the "Continuator" project and beyond, while seemingly more limited at first sight, is in fact much more appealing, because it clearly adopts the view of a reflexive human–robot interaction in musical composition.

88.

Generalized anthropogeny and dehumanizing. Walter Christensen Jr., "Does the Universe Have Cosmological Memory?," *Journal of Cosmology* 14 (2011). Ray Kurzweil, *The Age of Spiritual Machines.* Donna Haraway, *Simians, Cyborgs, and Women: The Reinvention of Nature* and *When Species Meet.* Cary Wolfe, *Before the Law: Humans and Other Animals in the Biopolitical Frame.*

89.

Posthumanism, posthumanities. N. Katherine Hayles, *How We Became Posthuman: Virtual Bodies in Cybernetics, Literature, and Informatics*; Pepperell, *The Posthuman Condition: Consciousness beyond the Brain*; Cary Wolfe, *What Is Posthumanism?*

90.

Prehistoric cave painting. André Leroi-Gourhan, *Préhistoire de l'art* and *L'art pariétal*; Georges Bataille, *Lascaux*; Randall White, *Dark Caves, Bright Visions: Life in Ice Age Europe* and *Prehistoric Art: The Symbolic Journey of Humankind*; Jean Clottes and David Lewis-Williams, *Les chamanes de la préhistoire.*

95.

Threshold of consciousness. Stanislas Dehaene, 2009 course at the Collège de France (see http://www.college-de-france/site/stanislas-dehaene/ and http://www.college-de-france/media/stanislas-dehane/UPL62003_Dehaene.pdf); Dehaene and Naccache, "Towards a Cognitive Neuroscience of Consciousness."

96.

Psychoanalysis and experimental psychology. Freud, "Project for a Scientific Psychology," in *Standard Edition of the Complete Psychological Works* (New York: W. W. Norton, 1976), vol. 1, 283–397. Lionel Naccache, *Le nouvel inconscient*; Georg Northoff, "Psychoanalysis and the Brain," *Frontiers in Psychology* 3 (2012): 71. K. Guenther, "Recasting Neuropsychiatry," *Psychoanalysis and History* 14-2 (2012): 203–26. Edmond Cros, *De Freud aux neurosciences.* Catherine Malabou, *The New Wounded: From Neurosis to Brain Damage.* Adrian Johnston and Catherine Malabou, *Self and Emotional Life: Philosophy, Psychoanalysis, and Neuroscience.* An interesting synthesis on the conscious and the unconscious in current cognitive neuroscience appears in Max Velmans, *Understanding Consciousness.*

97.

Qualia, etc. Thomas Nagel, "What Is It Like to Be a Bat?," *Mortal Questions,* chapter 12. Hilary Putnam, *Reason, Truth, and History*; John Searle, *The Mystery of Consciousness*; Daniel Dennett, *Consciousness Explained*; Michael Tye, "Qualia," *Stanford Encyclopedia of Philosophy,* http://plato.stanford.edu/entries/qualia/; Antonio Damasio, *The Feeling of What Happens: Body and Emotion in the Making of Consciousness*; Kirk, "Zombies," *Stanford Encyclopedia of Philosophy,* http://plato.stanford.edu/entries/zombies/. Alwyn Scott, "Physicalism, Chaos, and Reductionism" in *The Emerging Physics of Consciousness,* Tuszynski, ed., makes a good plea for an approach in terms of complex dynamic systems (rather than the quantum theory developed by Penrose and Hameroff). Stanislas Dehaene, 2009 course at the Collège de France.

101.

The paranormal. Alan Turing, "Computing Machinery and Intelligence," 453, about "extra-sensory perception": "These disturbing phenomena seem to deny all our normal scientific ideas. How we would like to discredit them! Unfortunately the statistical evidence, at least for telepathy, is overwhelming."

102.

New ideas on the soul. A notion of the soul as "informative" has been advanced in a religious perspective by Mark Graves, *Mind, Brain, and the Elusive Soul: Human Systems of Cognitive Science and Religion.* I like the idea of making use of cognitive science to revive old concepts, but of course, I do not share the author's prerequisite nor the theological consequences of his hypothesis. After writing this manuscript, I also read the excellent book by Philippe Lazar, *Court traité de l'âme.*

Index

(continued from page ii)

Laurent Dubreuil is professor of comparative literature and romance studies as well as a member of the Cognitive Science Program at Cornell University. The current editor of *diacritics*, he is the author of six books of literary theory and philosophy, including *The Empire of Language,* and several works of creative writing in French, including *Génération romantique.* His initial research in cognitive science has been supported by the New Directions program from the Andrew W. Mellon Foundation.